IMAGES
of America

WINDMILLS AND WATER MILLS OF LONG ISLAND

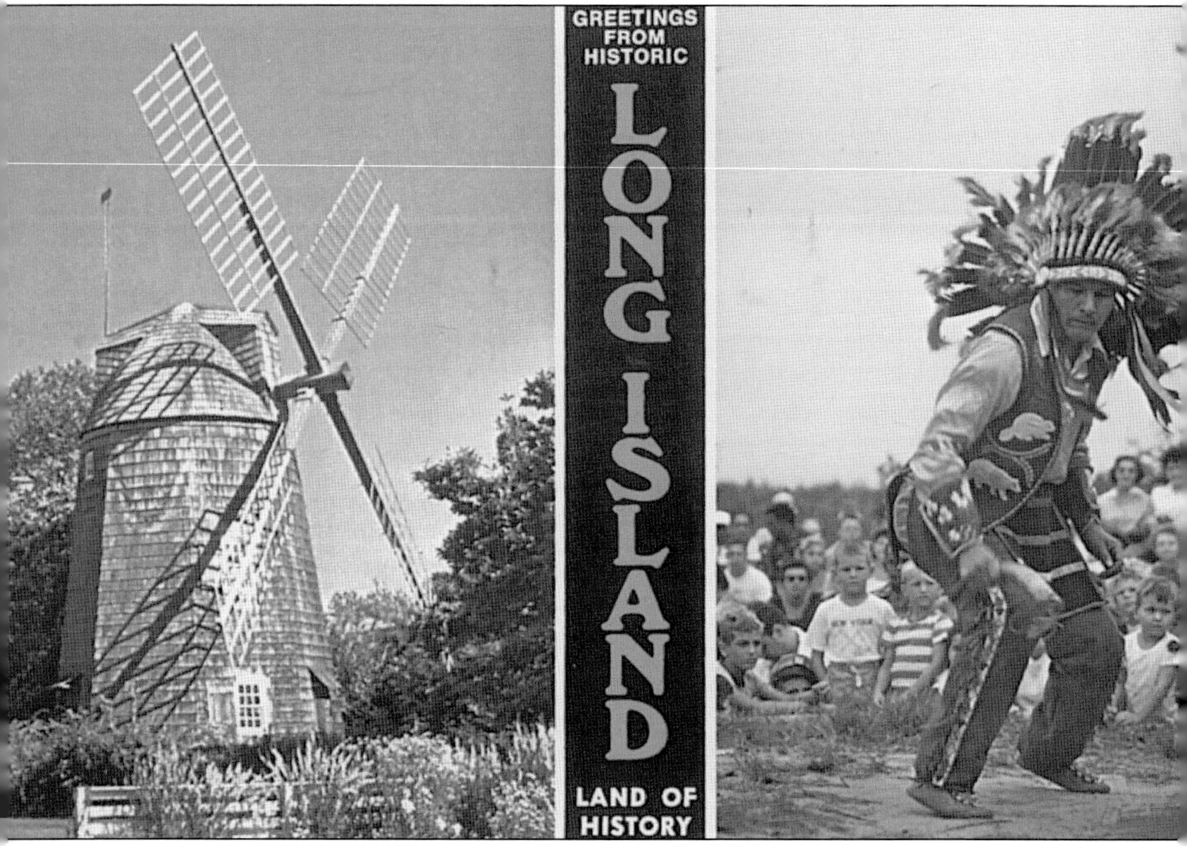

Chief Bright Canoe is performing the Indian Challenge Dance. Each Labor Day weekend native inhabitants of the Shinnecock Reservation in Southampton stage a powwow. Native Americans from other reservations around the country join in the annual event. Native products are sold, food is served, and the entertainment is attractive. The Native-American population on the island never exceeded 6,500. Tribal life ebbed out in the early 17th century. In 1649, white inhabitants began to settle Long Island. They brought their culture, customs, and way of life, which included their windmills for grinding grain. This is the Patango Windmill in Easthampton.

IMAGES
of America

Windmills and Water Mills of Long Island

Sr. Anne Frances Pulling
and Gerald A. Leeds

ARCADIA
PUBLISHING

ISBN 978-0-7385-0288-5

Published by Arcadia Publishing
Charleston, South Carolina

Printed in the United States of America

Library of Congress Catalog Card Number: Applied For

For all general information contact Arcadia Publishing at:
Telephone 843-853-2070
Fax 843-853-0044
E-mail sales@arcadiapublishing.com
For customer service and orders:
Toll-Free 1-888-313-2665

Visit us on the Internet at www.arcadiapublishing.com

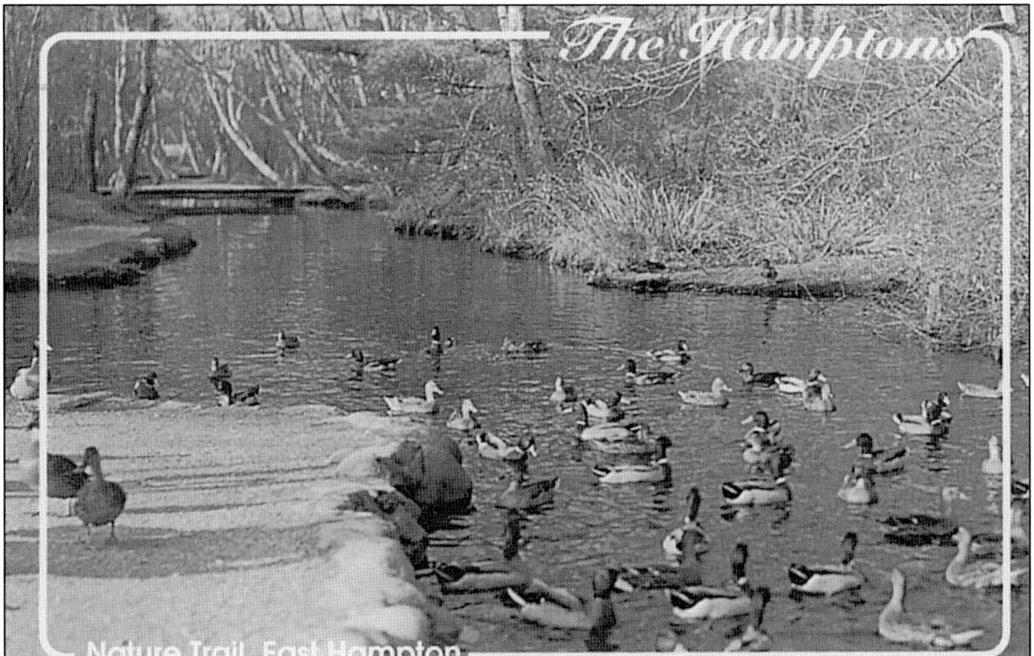

This picturesque pond is indicative of many quiet, serene, and scenic localities on Long Island. The mallard duck, a native of the Northern Hemisphere, is attracted to the numerous creeks, marshes, ponds, and lakes the island offers. These are favorite gathering places for generations of children who visit and feed the ducks. Unlike the white Peking ducks raised on eastern Long Island, the mallards are colorful and graceful. They are quick to respond if they fancy a treat.

Contents

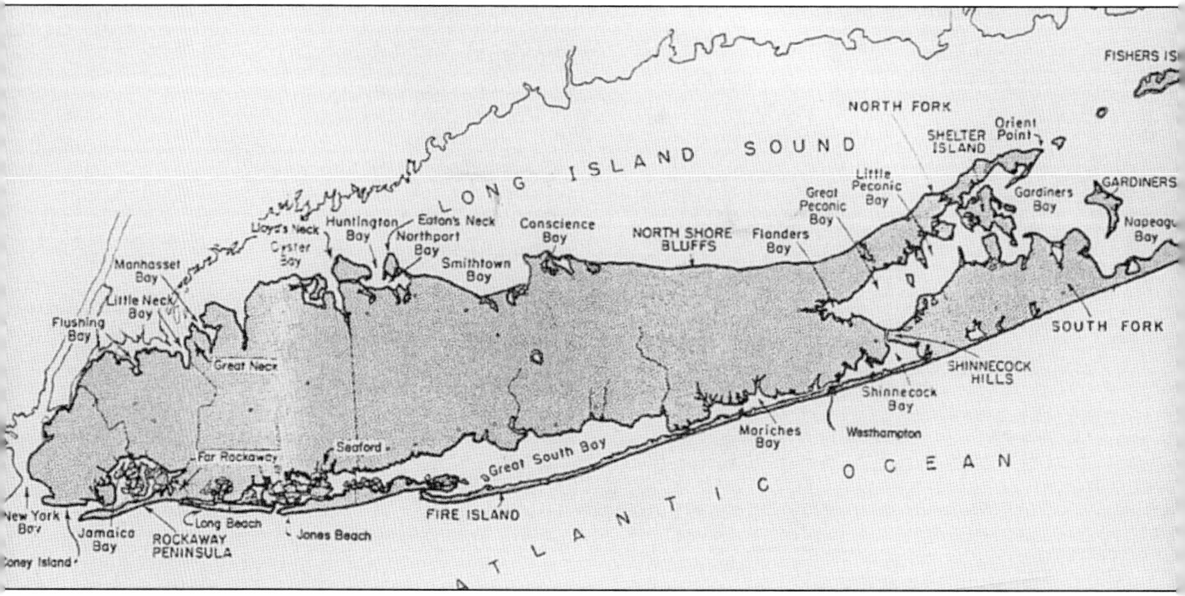

Long Island is a 125-mile stretch of land jutting into the sea off New York Bay. It is 20 miles at its widest point, from Seaford to Huntington. Native Americans called it "Paumonak" or "The Land of Tribute." The tail forms two peninsulas on the east end known as the north and south forks. The shape of Long Island was sculptured through the action of glaciers. The island is divided into four counties: Kings or Brooklyn, Queens, Nassau, and Suffolk.

ACKNOWLEDGMENTS

The authors would like acknowledge the generous help of Bob Cammann, Stan Ziminiski, Howie Friedman, and Kyle Berris, who helped find the many mills on Long Island as well as sharing photos, slides, postcards, and articles on the mills. A special thanks goes to Stan and Martha Kellner for suggesting that the authors write a book. The Long Island libraries, historical societies, The Society for the Preservation of Old Mills (SPOOM), The International Moninological Society (TIMS), The Society for the Preservation of Long Island Antiquities (SPILIA), the Long Island Studies Institute at Hofstra University, and many mill friends were all extremely helpful and generous with their time and resources. A special thank you to Jim Owens, Tom Glick, Tom Stock, Dave Free, John O'Connor, Len Minerva, and Town Historian David Overton in Brookhaven for their invaluable support.

Prayerful gratitude goes to all who helped by supplying photographs, assisting with caption material, and proofreading. A special thanks to the Sisters of Mercy of the Dallas Regional Community, for their support and encouragement in this project.

INTRODUCTION

Windmills are the wooden technology of yesteryear. These quaint structures are picturesque vestiges of an almost legendary era a century ago. Mills symbolize clean, efficient, cost-free technology. They need no fuel and cause no pollution. They employ free resources to grind grain, pump water, and generate electricity. Mills represent the oldest example of American craftsmanship in the wooden age.

Mills are viewed as artistic, graceful, elegant, and charming. We capture them in song, story, and on canvas. They are nostalgic landmarks of the past, but early settlers viewed mills as structures critical to their needs and the economic well being of the region. The mill was a necessity. They accomplished tasks essential to survival. The miller depended upon it for his livelihood and the general population for flour and daily bread.

The flat topography of Long Island and the steady winds from the bays and the Atlantic Ocean set the stage for windmills. They became an ideal source of dependable power for the early inhabitants of eastern Long Island.

Mills were usually built on stone foundations. The millers would carry bags of grain up the steep, spiral, uneven steps to the third floor of the mills, where the grain was placed in a hopper. It then flowed into wooden chutes, which led to the millstones below. Here the grinding process took place. Once ground it was poured from a spout under the stones unto a belt that took it to the bags. Each farmer's grain was ground into his own bag of closely woven homespun linen.

The sound of the mill was music to the ears of the miller. Once set into action the noise and motion began with a little creak and shudder. This rapidly accelerated to a great crescendo. The sails creaked and swished in the wind. They swept majestically round as the runner stone revolved, just above the stationary bed stone, with a rumbling, roaring sound that filled the dusty air.

The graceful sails then accelerated rapidly, sometimes traveling at 30 miles per hour. They vibrated the entire building; every nook and cranny of the mill quivered into life. Vibrations gripped the structure as the wind propelled the sails. The mill clattered rhythmically as sacks of grain were taken to the top and began their journey down to the millstones.

The south fork of eastern Long Island contains the greatest number of surviving windmills in the United States. They dot the landscape as sentinels of the past. This area provides a rare opportunity to investigate the evolution of technology in a distinct region. Most of the surviving windmills are smock mills, which means they have a stationary octagonal tower surmounted by a cap holding the sails. The cap revolves to keep the sails facing the wind. Wind-driven mills existed in Persia as early as the 7th century. European mills appeared in France and England in the 12th century. These early wooden structures were called post mills. They rotated manually around a central post to bring the sails into the wind. Tower mills developed in the 4th century in France. They comprised stationary machinery topped by a rotatable cap bearing the wind shaft and gearing. The tower was usually octagonal and built of wood.

Mills were adapted to a variety of uses such as pumping water, sawing wood, making paper, pressing oil from seeds, and grinding different materials. In 1745, the fantail was a major innovation because it automatically rotated the sails into the wind.

Water mills were prevalent throughout Long Island before 1700. Some of them were powered by the tide. Gristmills and sawmills were often powered by the same wheel. Many communities had at least one tide-powered mill.

Millers developed an affection for their mills and machinery, as they spent most of their waking hours working. They knew its strengths, weaknesses, and its idiosyncrasies. They were well aware that each mill has its own individual character and personality. Millers frequently referred to their mills as "she" and "her."

By the turn of the 20th century, processed flour was shipped to eastern Long Island from the city, thereby reducing the need for mills. Local farmers began cultivating potatoes, duck farms sprang up, and farming was revolutionized. Long Island rapidly became populated, the Gold Coast era dawned, and many estates fringed both north and south shores. Long Island became known as the "Cradle of Aviation." Pioneering flying ventures took place on its flat lands and aircraft manufacturing emerged.

Long Island has always been associated with innovation and progress. It was the home of two signers of the Declaration of Independence (William Floyd of Mastic and Francis Lewis of Queens), and the world's first toll road, Motor Parkway, was constructed from Queens to Lake Ronkonkoma by William Vanderbilt in 1908. Roosevelt Field was the point from which Charles Lindberg took off on his historic flight to Paris in 1927. The nation's first supermarket, King Kullen, was established in Great Neck in 1930, and the Luna Module, which carried the first man to the moon in July 1969, was built at the Grumman Aircraft Factory in Bethpage.

The new millennium will bring further advances in science and technology. Let us remember the innovations of our Long Island ancestors and retain the mills as a vital part of Americana. These picturesque remnants of the past, now in a nostalgic pasture, deserve to be reserved as national treasures. Long Island proudly claims a large share of these sentinels of yesteryear on the south fork's historic east end.

One

Sentinels of Yesteryear

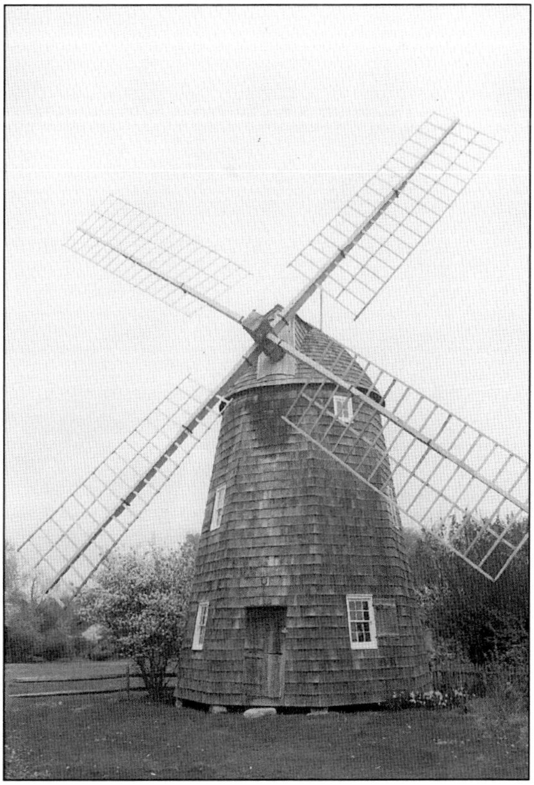

Samuel Schellinger built Pantigo Mill on Mill Hill in 1804. When ownership changed the mill was moved to the corner of Pantigo Road and Egypt Lane, where it operated successfully for seven decades. In 1917 it was moved to the 17th-century house Home Sweet Home. The village undertook intensive repairs on the mill, which is part of the Easthampton Museums. Native Americans and early white settlers lived peaceably on the island for many years.

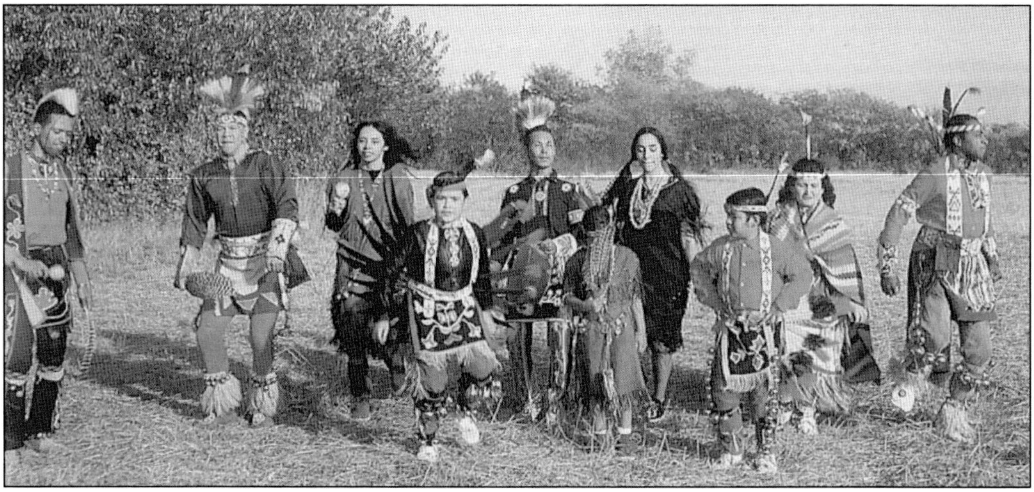

Native Americans representing various tribes perform a ceremonial dance while visiting the outpost on the Shinnecock Reservation in Southampton. They were united in an island-wide confederation, each with its own territory and chief. They called the island "Paumonak," meaning "land of tribute," since they had to pay tribute to contemporaries in upstate New York. Through the years Long Island farmers have discovered and collected artifacts that reveal a high degree of Native-American artistry.

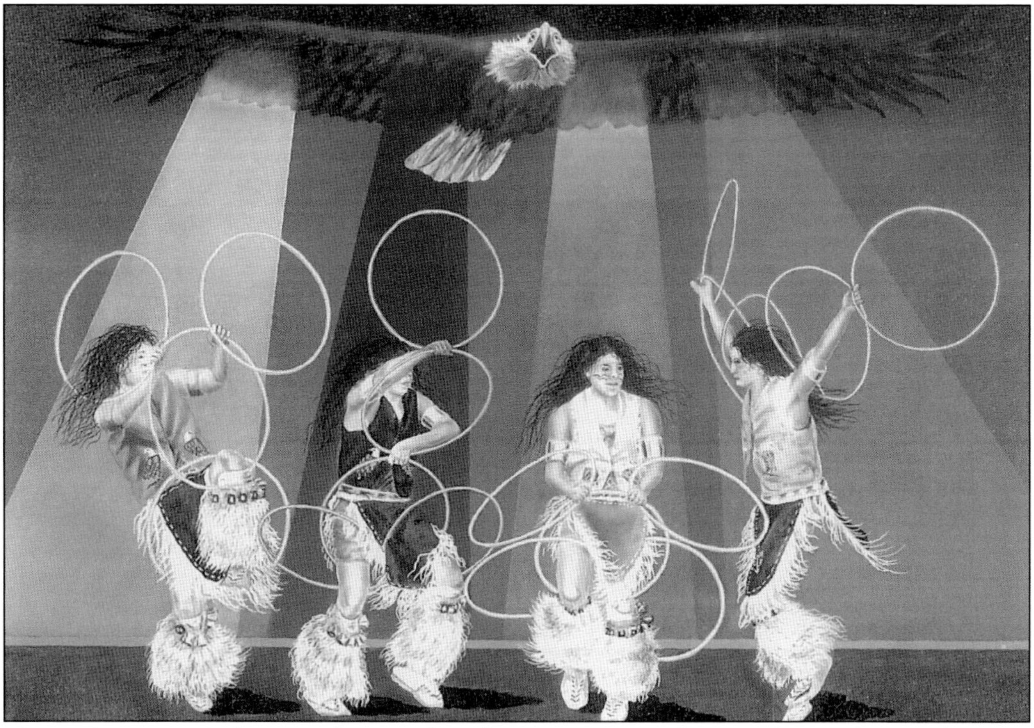

These American Indians perform a four-direction prayer dance. The mighty eagle, as a messenger from the heavens, radiates the power of the four directions to the Hoop Dancers. East, west, north, and south bless the dancers as they move gracefully like the wind. There are two Native-American reservations on Long Island—the Poosepatuck Reservation in Mastic and the Shinnecock Reservation in Southampton, where the powwow is performed each year.

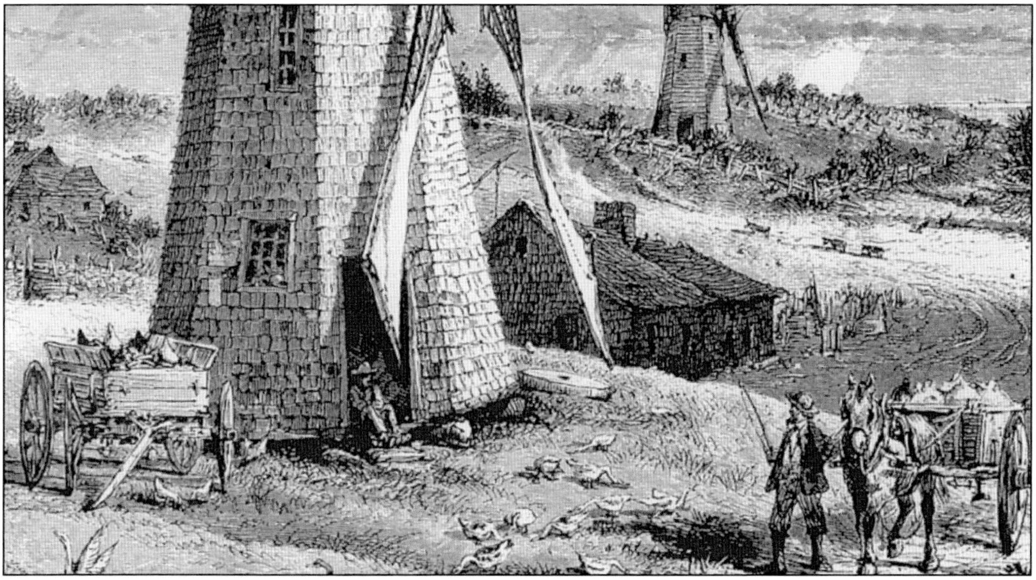

Easthampton, originally called Maidstone, has retained its Colonial character. Many early settlers came from County Kent around Maidstone, England. Easthampton is a mecca for the preservation of local history. Hook Mill and Pantigo Mills were the hub of activity. Note the miller awaiting the farmer on the steps of Hook Mill. Grain was delivered daily to the mills in 1870. The grinding of corn and wheat were fine examples of wooden technology.

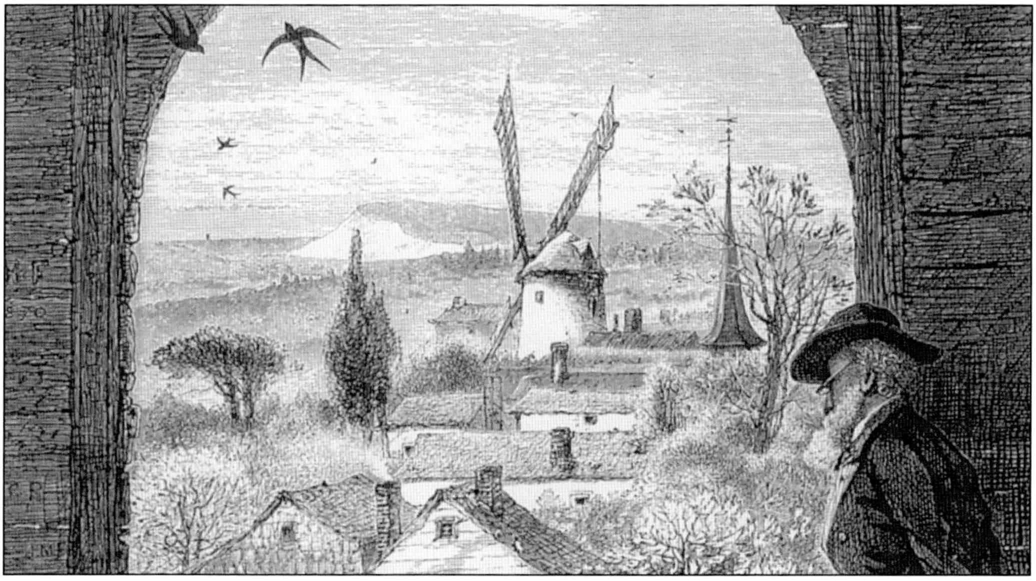

Rev. Lyman Beecher, father of Harriet Beecher Stowe, enjoys a bird's-eye view of Easthampton from the belfry of his Presbyterian church in 1870. The original settlement of 1649 was on the broad commons now part of the historic district. There were dwellings on each farm with huge fields. The town was occupied by the British during the Revolution. The little house of worship was built in 1717.

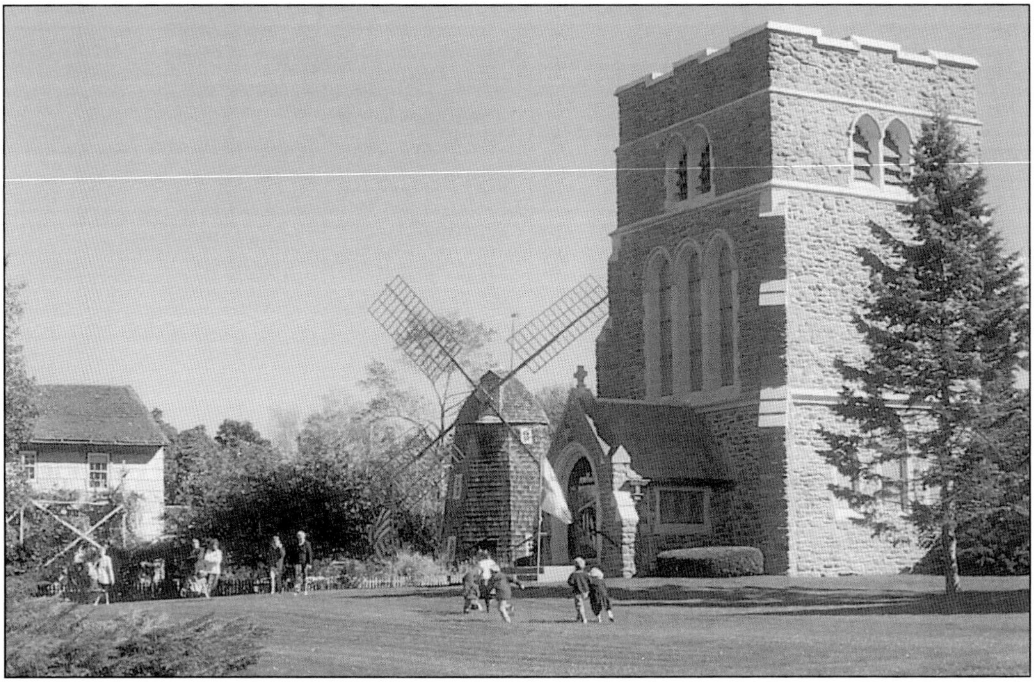

Pantigo Mill was rebuilt for Abraham Gardiner, who sold his share when the Gardiner Mill was built. Originally, it stood on Village Green. In 1850, Felix Dominy purchased it and moved it to the Pantigo Amagansett Road. In 1916, Gustave Buek bought it and moved it to Home Sweet Home, where he had it repaired and opened as a museum. The Episcopal Church of St. Luke is in the foreground.

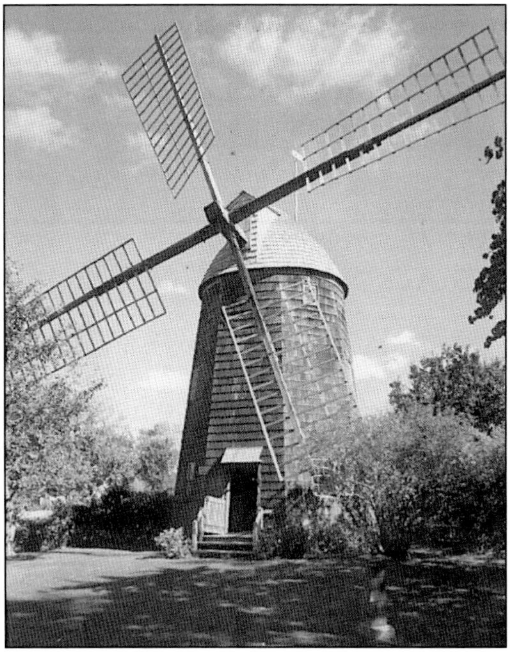

Pantigo was once Hunting Miller's Mill. A story goes that Miller was in church when the minister heard the wind and announced "Church will be dismissed, Miller, go to your mill and grind." The law was that the minister's grist should be ground first.

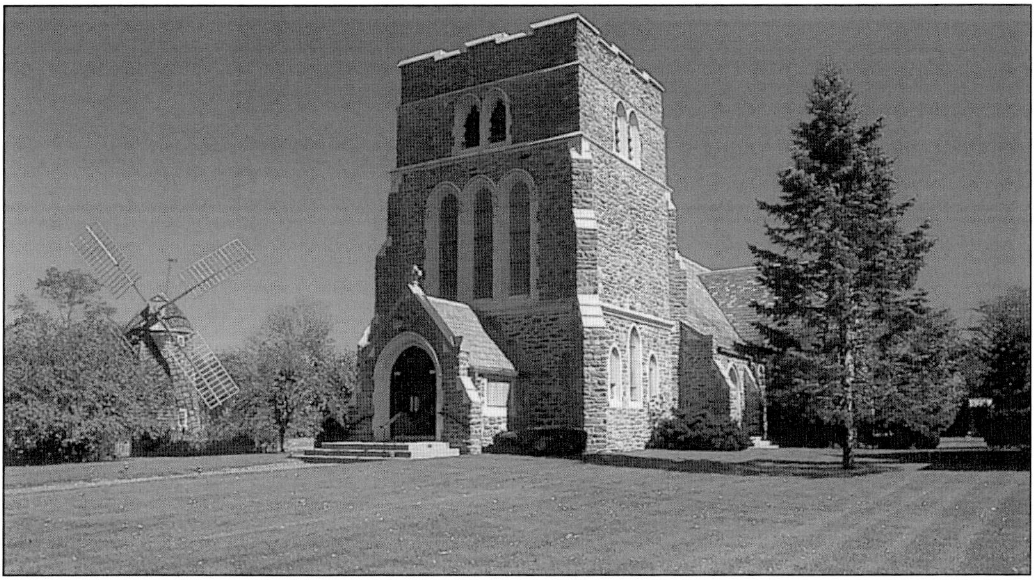

The Episcopal Church of St. Luke is on the grounds of Home Sweet Home. In 1907–1908, this solid-looking building was constructed. The stone was obtained from the blasting of tunnels in New York City. It is porous and requires constant repair. The church is a copy of St. Mary's in Ashford, Kent, just outside Maidstone, England. An Easthampton physician suggested the name.

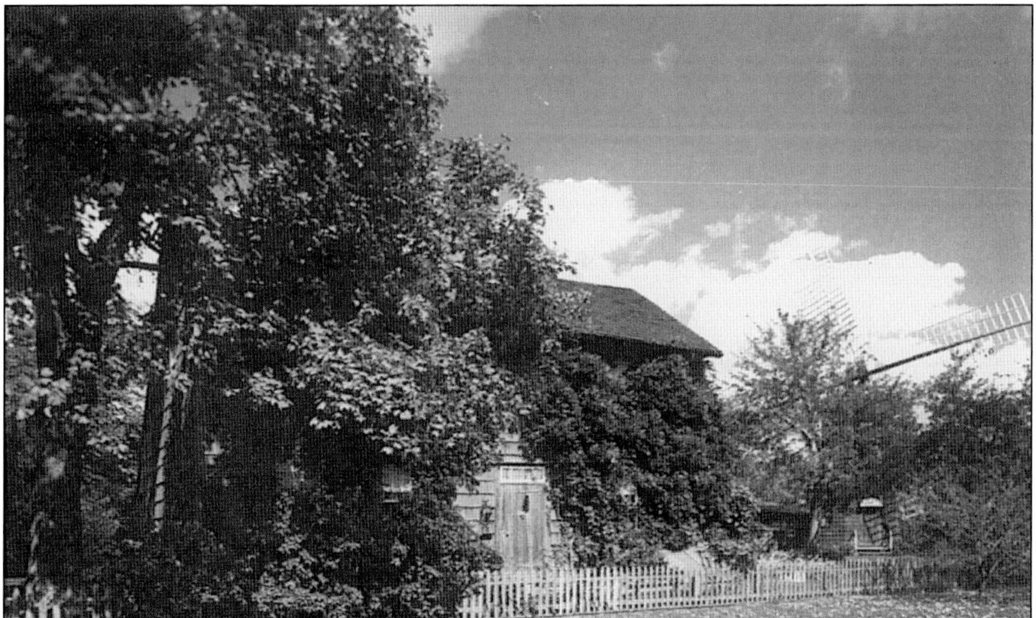

The Pantigo Mill and Home Sweet Home are nestled in a panorama of rustic beauty. Windmills gave rise to some of our familiar expressions. "The same old grind" compares lack of diversity, "milling around" indicates going around in circles like the millstone. "Run of the mill" denotes anything average. "Put through the mill" designates difficult experiences and "go with the flow" indicates joining the crowd.

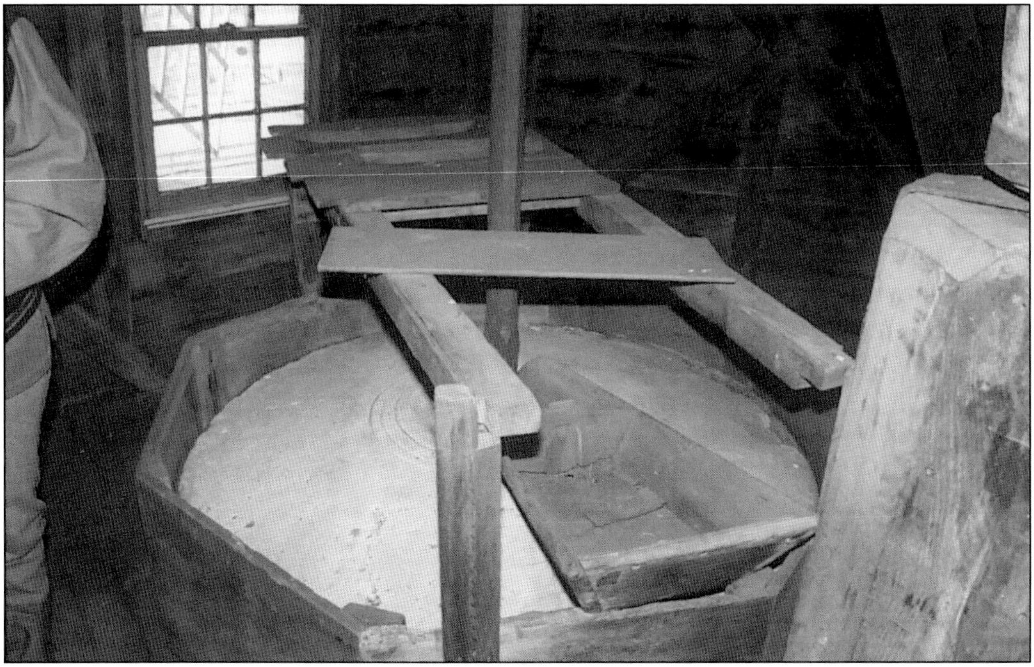

The huge millstone was a necessity in the grinding process. The mill has a pair of large, circular stones between which grain is ground. The turning cap stone (or upper stone) does the grinding; the bed stone (or bottom stone) does not move. Revolving gears cause the millstone to rotate. The Pantigo Mill has double-sided sails with bars on long stocks mortised through the wind shaft.

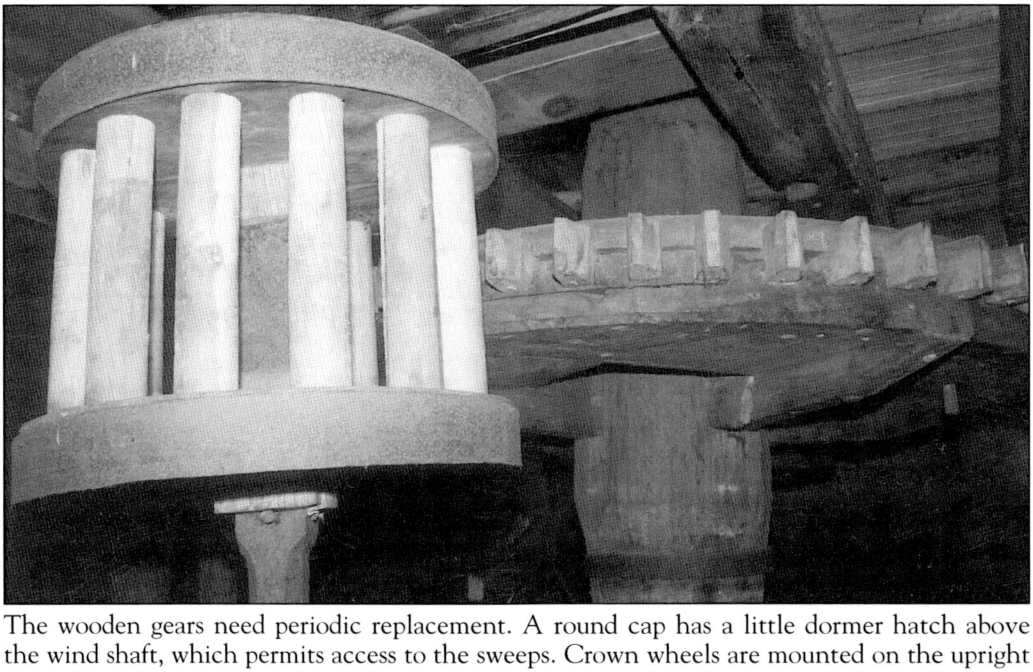

The wooden gears need periodic replacement. A round cap has a little dormer hatch above the wind shaft, which permits access to the sweeps. Crown wheels are mounted on the upright shaft above the great spur wheel. The pegged gear superseded the lower crown wheel. The upper crown wheel drives a layshaft that powers the bolter.

The Gardiner Mill, built in 1884, stood on the Easthampton Commons for many years, minus its sails. It is a smock mill with conical cap, gabled dormers, and double-sided sails. They had a very pronounced twist. Gardiner Mill stands on its original site near Maidstone Lane on the east side of Main Street. Robert David Lion Gardiner, a descendant of Lyon Gardiner, now owns the mill.

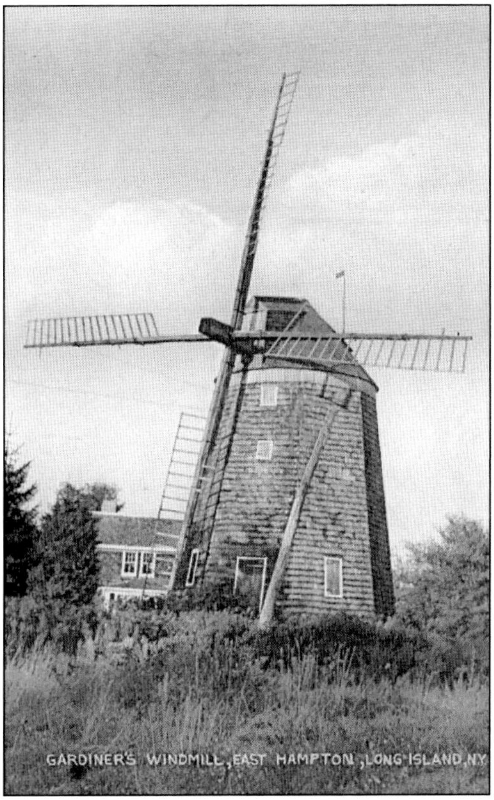

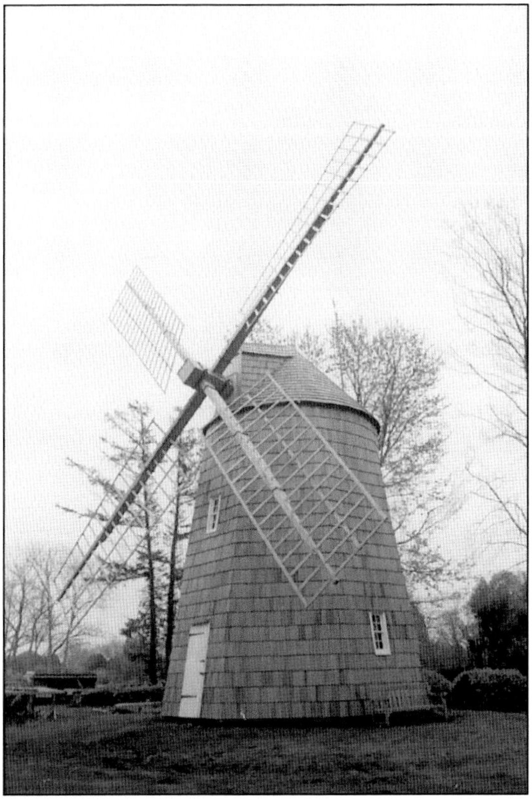

Gardiner's Mill was constructed with a new type of intermediate gearing to drive two pairs of millstones. When it was planned, John Lion Gardiner became a shareholder. Syndicates owned it and is the only surviving mill with millstones driven from below. The sails were lost over the years. The Gardiner Windmill on the Easthampton Commons was restored and given new sails.

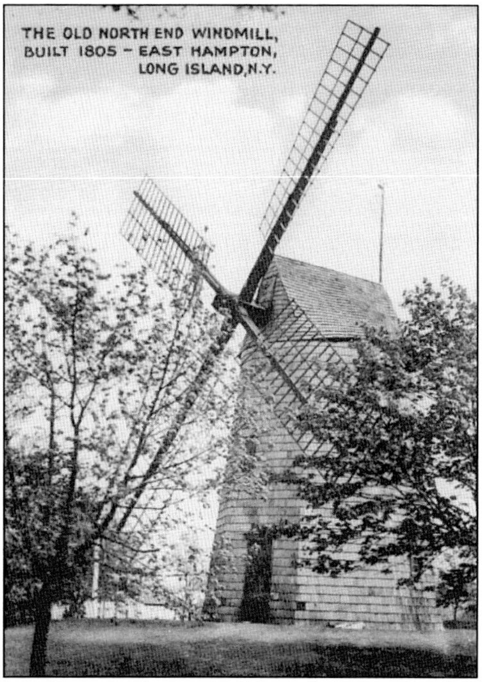

Hook Windmill is also called the old North End Mill because of its original location. It retains more internal machinery than the nationally significant Long Island Windmills. It was repaired in 1939 after lying idle for three decades. The grinding of corn and wheat are fine examples of wooden technology. Hook Mill is the first building on the south fork of Long Island to be preserved as an historical landmark.

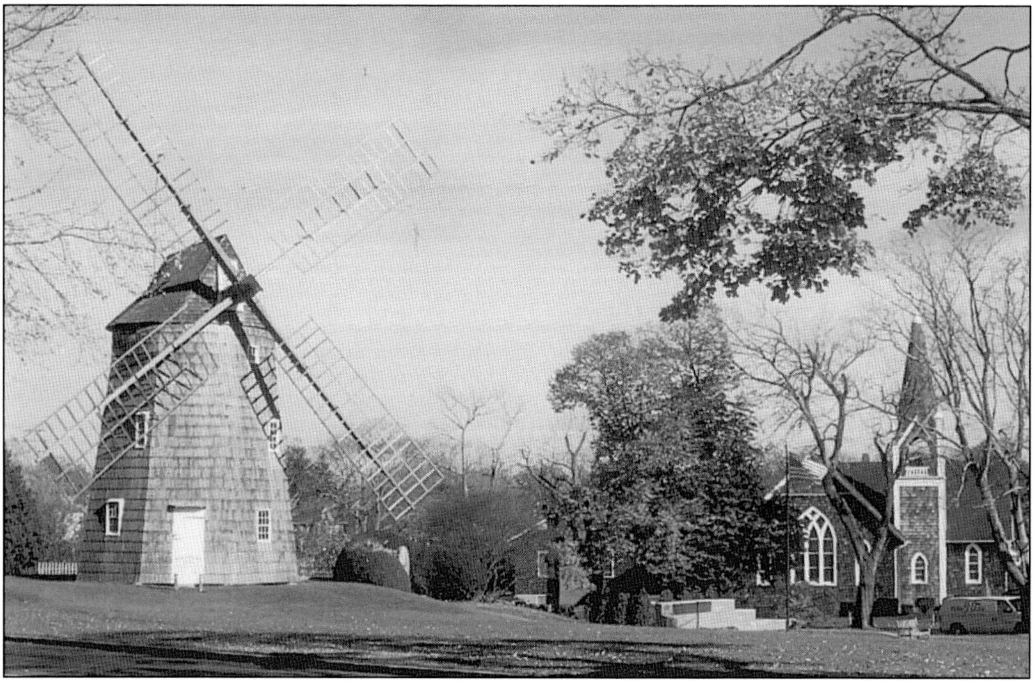

Hook Mill was an excellent example of Dominy's skill as a millwright. A system of wooden gears power two pairs of millstones, which revolve the cap and power auxiliary machines to process grain. Dominy installed time saving devices such as a sack hoist, a grain elevator, a screen to clean the grain, and bolters to sift the cornmeal and flour. Nathaniel Dominy brought the Hook Mill, with all its wooden components, to its highest degree of sophistication. It stands near the United Methodist Church of Easthampton.

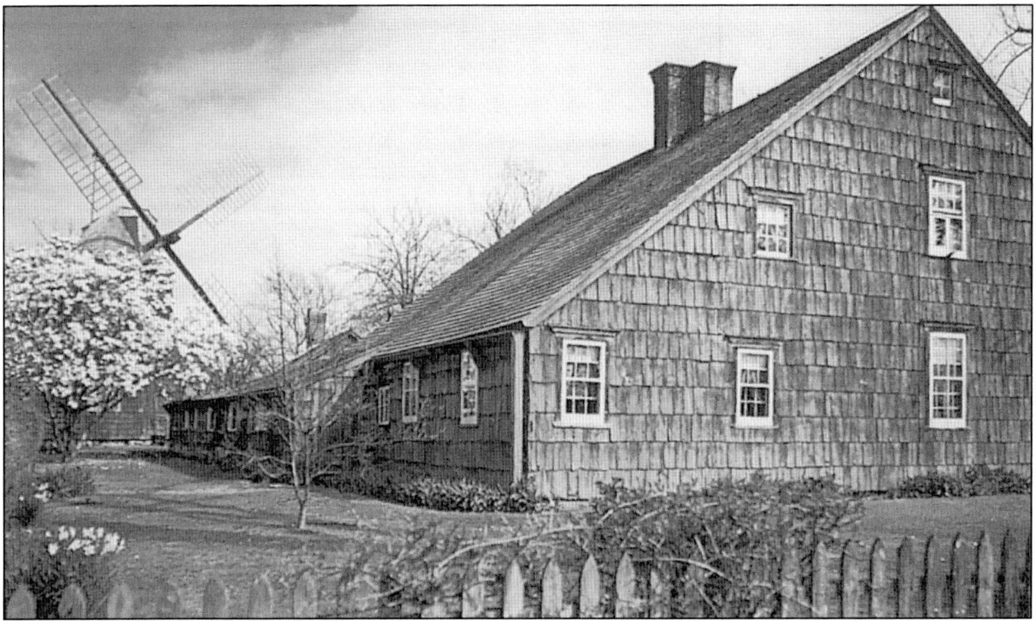

John Howard Payne, author of the world famous song "Home Sweet Home," was born in Easthampton on June 9, 1791, and spent much of his boyhood here. "Home Sweet Home" was first sung in his opera *Clari, The Maid of Milan* at the Royal Theatre in London on May 8, 1823. Payne died as an American Consul in Tunis, Africa on April 9, 1852. He derived inspiration for the famous song from his boyhood home in Easthampton. The home of Payne's immortal song preserves significant detailing of early American construction.

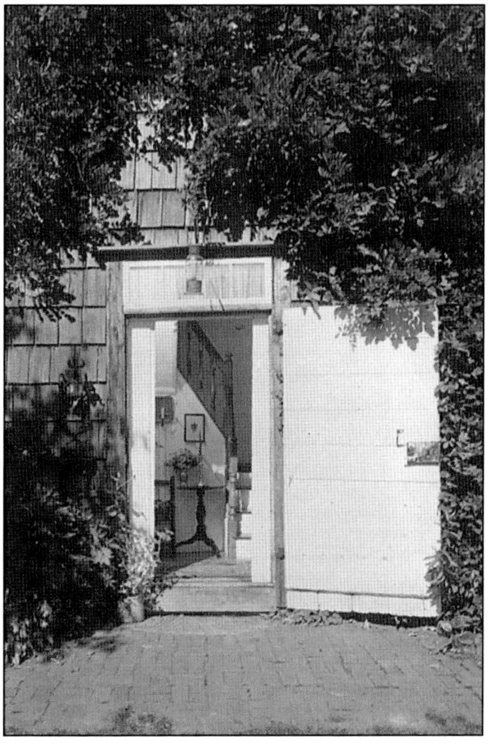

The front entrance hall at Home Sweet Home has remained much the same as when Payne lived here. It features a hand-worked stair railing, turned spindles, and detailed ornamentation of the early American period. In Payne's time, his grandfather, Aaron Issacs, who was president of Clinton Academy, owned the house. The village of Easthampton now preserves the historic edifice as a museum.

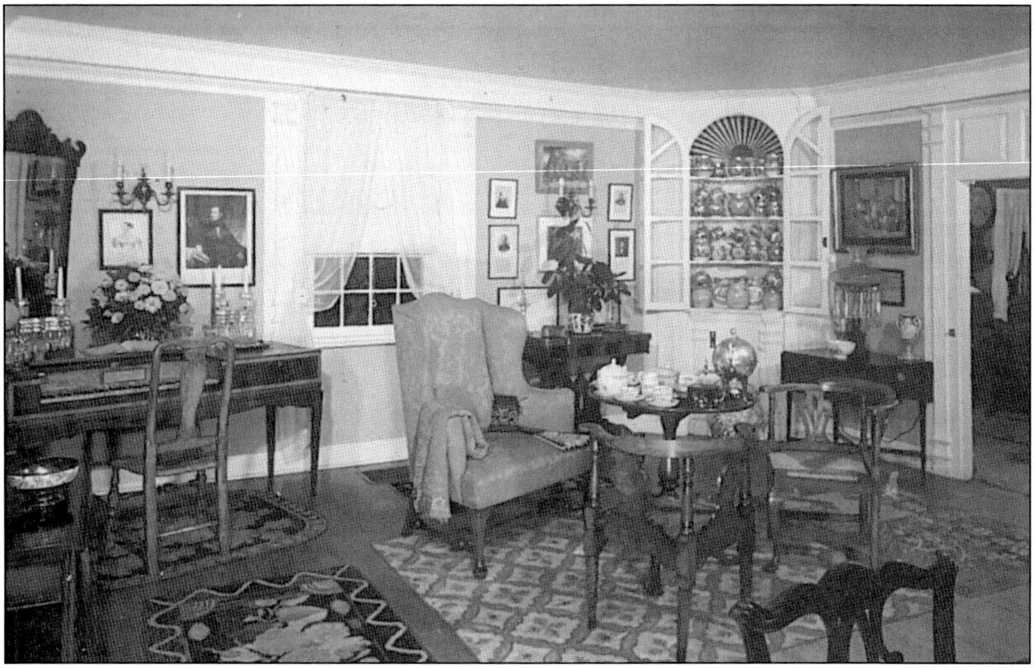

The living room at Home Sweet Home features English lusterware that preceded the silverware we know today. The lamps and candleholders are imported crystal. Detailed woodcarvings of the 18th century adorn the table, chairs, and desk. Note the lusterware in the cupboard. It was also made into useful containers and often stored where it would brighten dark corners. Staffordshire china is used on the table.

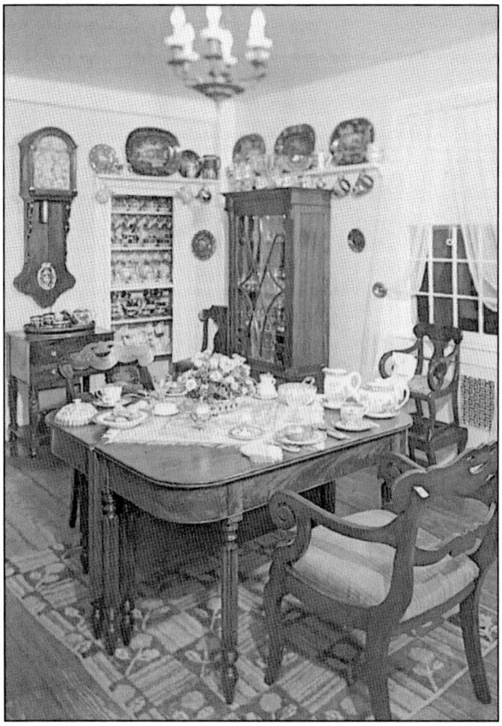

The dining room at Home Sweet Home is adorned in the American Empire decor of 1810. Delft china, lusterware, imported crystal, and Staffordshire china comprise the setting complete with scones. The historic edifice contains a notable collection of colonial antiques and John Howard Payne relics. Included is a silver jug that dates back to 1810. Note the detail in the woodcarvings on the furniture.

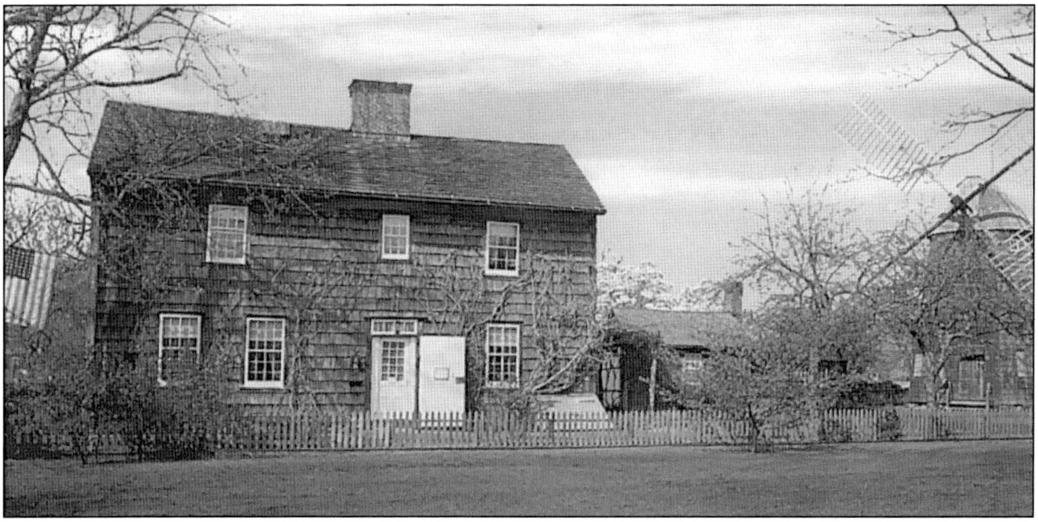

A front view of Home Sweet Home shows Pantigo Mill to the right. Home Sweet Home is a transitional, lean-to house that preserves significant detailing of early American construction. It is a 17th-century-shingled building with a typical saltbox roof and walls of white cedar. The shingles on the roof are three feet long. Restoration required custom-made replacements.

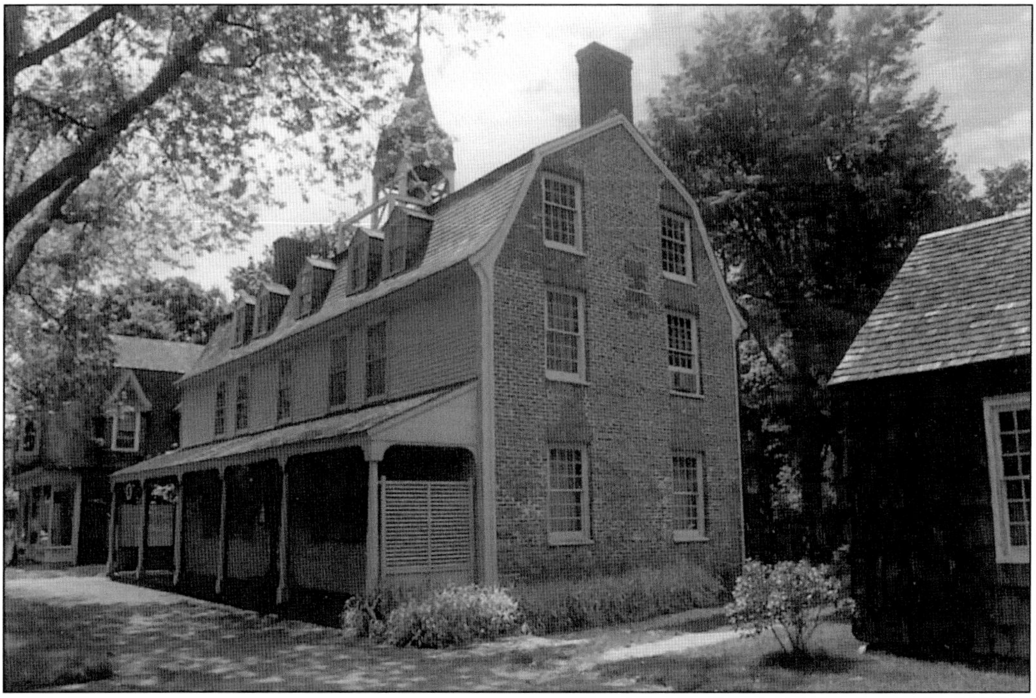

Clinton Academy, built in 1784, was the first chartered academy in New York State. Boys prepared for college here and it was a "finishing school" for girls. Co-education was unheard of! William Payne was the headmaster. The school drew students from as far away as the West Indies. It was established through the leadership of Rev. Samuel Buell, minister of the Presbyterian Church. He brought his flock through the Revolution and managed to establish the Academy.

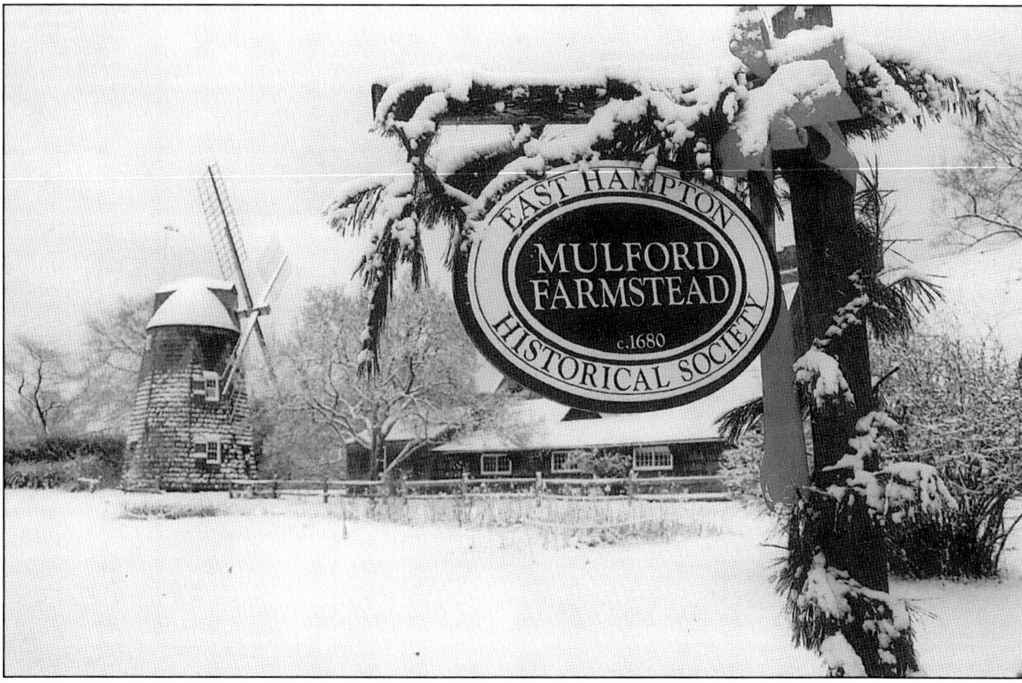

Mulford Farm has graced the Easthampton commons for 300 years. The farmstead dates back to 1680, when the Mulford family was among the earliest settlers. It is situated on four acres of land that have been preserved as one of the most intact 17th-century farms on eastern Long Island. Taken over by the Easthampton Historical Society, it offers a glimpse into farm life three centuries ago.

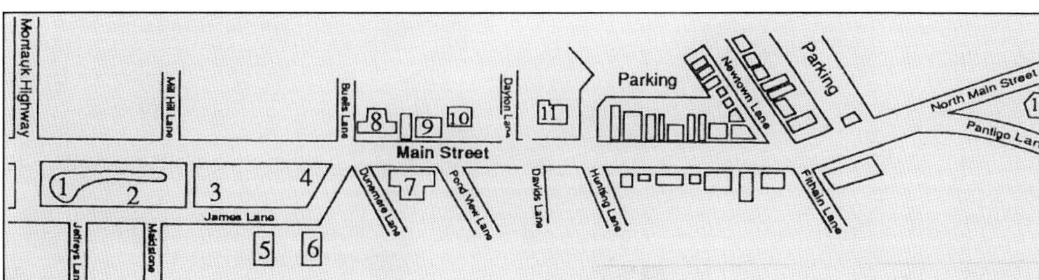

This map of East Hampton indicates what is where in one of America's most historic villages: 1] Town pond; 2] South End Cemetery; 3] Village Green; 4] flag pole; 5] Home Sweet Home; 6] Mulford Farm; 7] Guild Hall; 8] library; 9] Clinton Academy; 10] Town House; 11] Osborn-Jackson House; and 12] Hook Mill.

Hay Ground Mill was built in 1809 and stands on Windmill Lane in East Hampton south of Montauk Highway. Timber for the mill was cut in the northwest woods between Sag Harbor and East Hampton. It was last worked in 1919. All the millers have been descendants of original families. The mill still contains much of the original machinery and is used as a tea room.

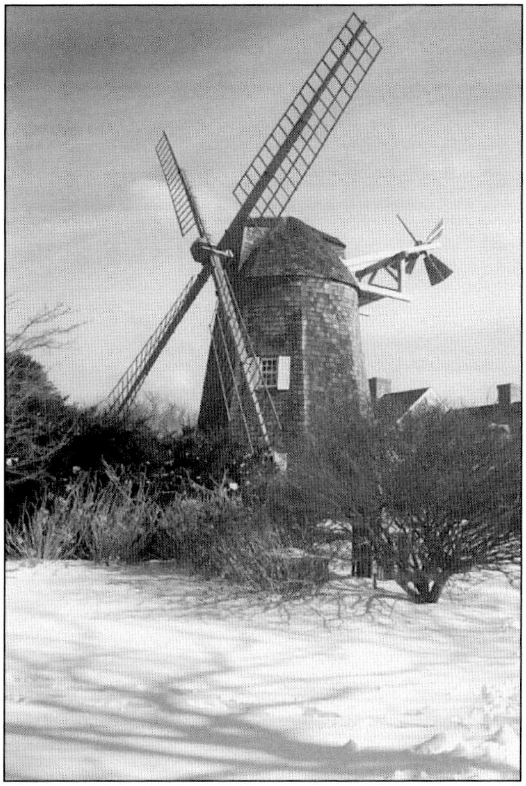

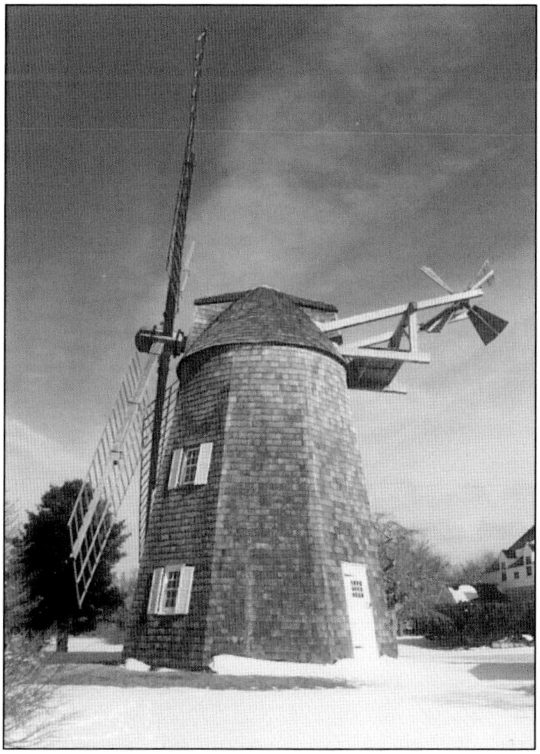

Hayground Mill was not constructed by Dominy V. There are too many features different from those Dominy built. In 1870, according to the *Census of Industry*, the mill was opened only four months of the year. It ground eight hundred bushels of wheat, four hundred of corn and eight hundred of oats in one season. In 1950, Robert Dowling purchased the mill. He moved it by rolling it along the beach to his estate on the dunes in East Hampton.

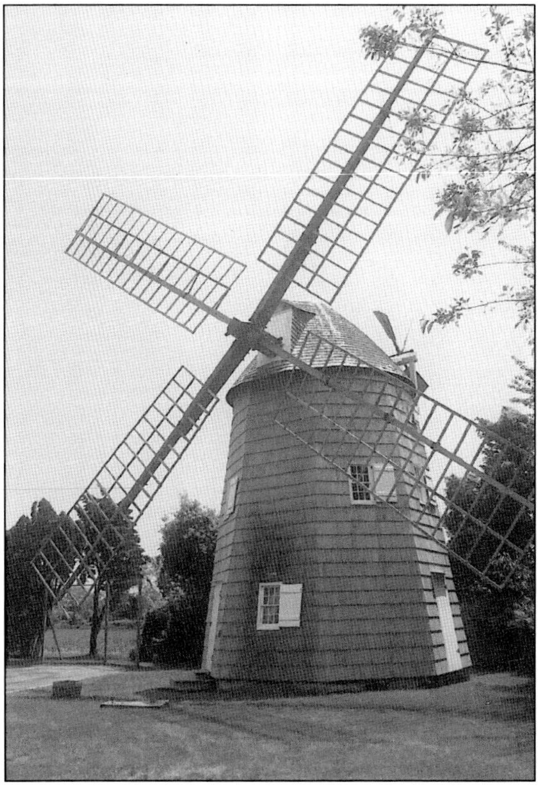

Wainscott Windmill was built in Southampton in 1813 to replace one that was burned. Purchased in 1840, it was relocated to a site beside Mill Hill Windmill. Richard Dunster was the only miller for both mills, which operated side by side. One ground wheat and the other corn. In 1858, the mill was moved to Wainscott where it operated for a half century. By 1860, the mill had ground 5,000 bushels of grain and produced $5,500 worth of flour and feed.

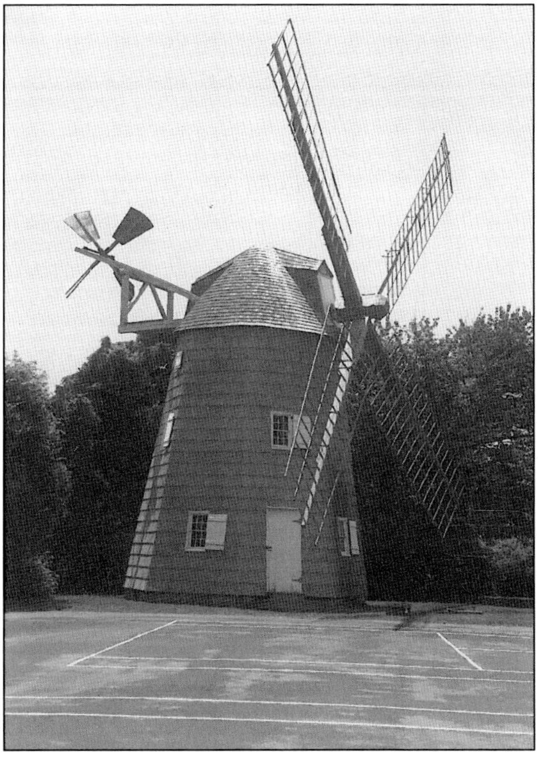

In 1912, the Wainscott Windmill, with sails intact, became the Wainscott Public Library. In 1922 Lathop Brown bought it and moved it to Montauk, just west of the lighthouse where it became part of a cottage he was building on the cliffs. By 1940, the federal government acquired the Montauk property. Brown gave the windmill to the Georgica Association, a private residential association. They removed the mill to their grounds.

Amagansett Windmill, was a post grist mill built in Miller Place by Samuel Schellinger of Amagansett in 1815. The Amagansett Mill Corporation moved it in 1829 to the west side of Windmill Lane behind the Methodist Church. It was later sold to Abraham Parsons, who operated it until his death. His sons kept the sails spinning until 1918.

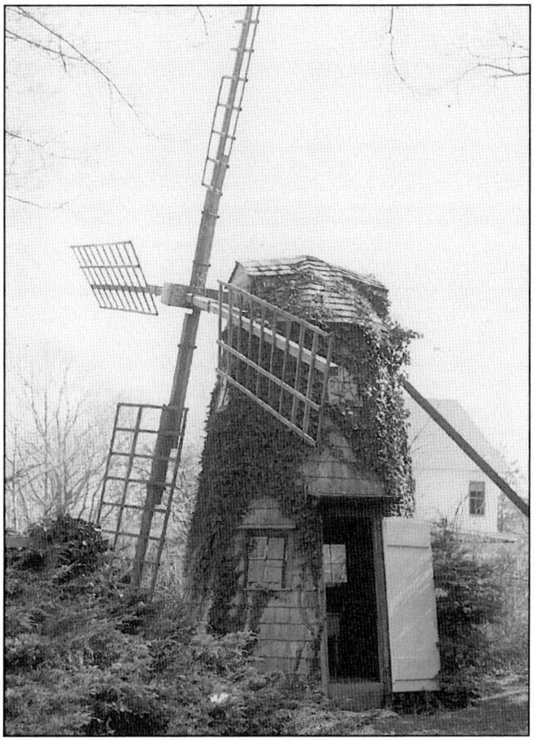

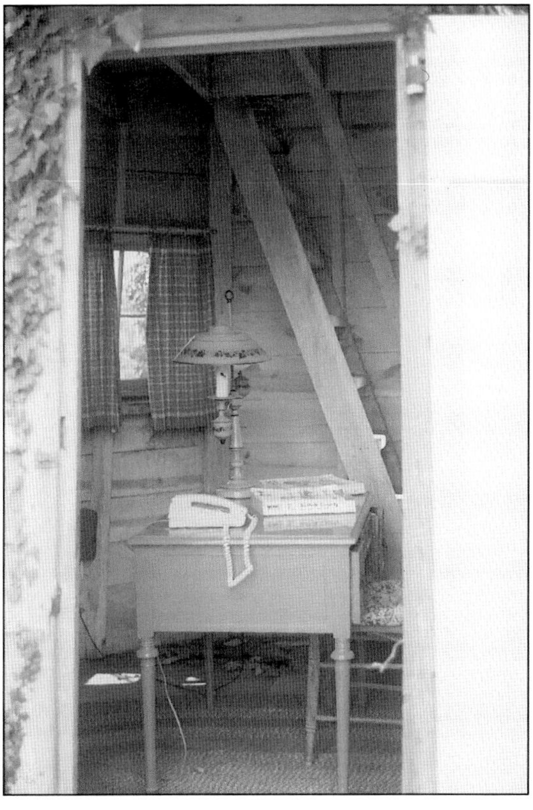

The body of the Amagansett post windmill revolved on a post. It is the earliest type of European mill and the interior mechanism was simple. The shaft to which the arms were fastened terminated in a huge crank. The outer ends of the arms were fastened in place by wooden pins to furnish power in a moderate breeze. The original mill was demolished by fire in 1924. A replica was constructed on Windmill Lane. A quaint office was established in the mill.

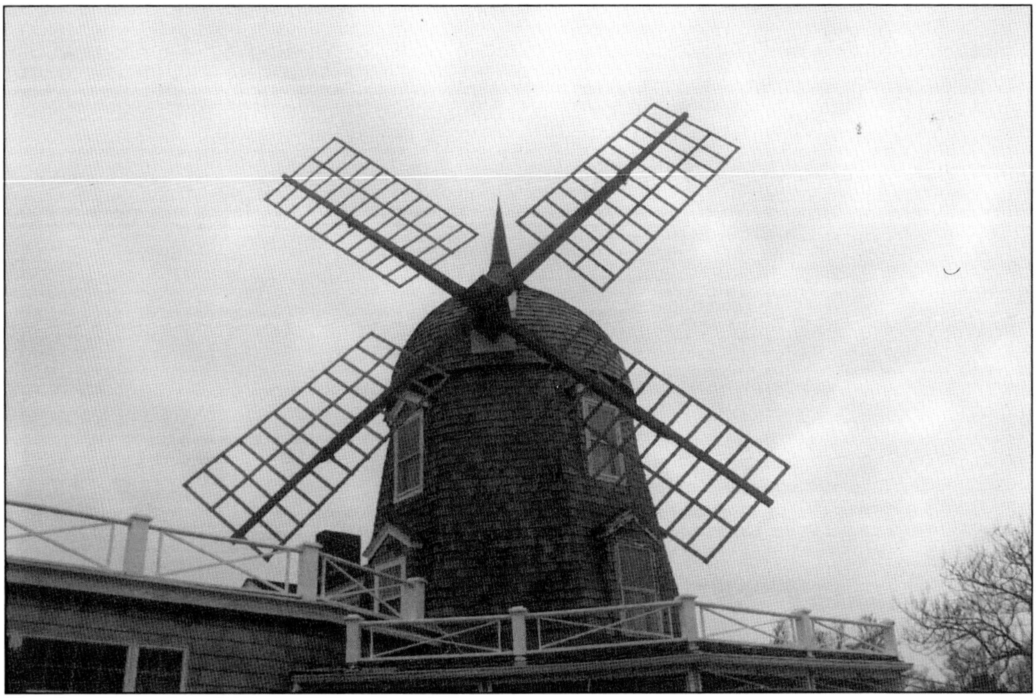

Good Ground Mill in Southampton was built on Shelter Island in 1807 for the Sylvester Dering. It had an internal winding mechanism. The Greek Revival architecture of the cap sets it apart from other mills. The cap rack is visible above the mill cottage. It was moved to Good Ground, now Hampton Bays, in 1806. Matthew Van Buren Squires owned the mill when it stood on Montauk Highway and Mill Corner.

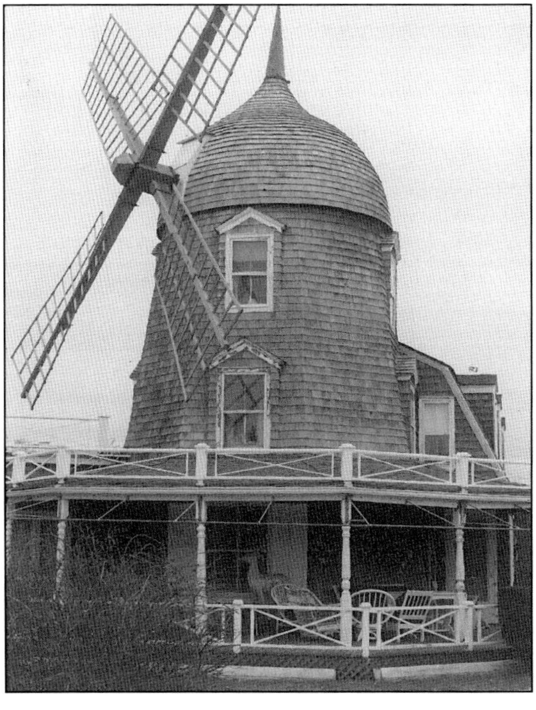

The Good Ground Mill operated until C.W. Betts moved it to Southampton in 1890. It became a cottage and stands in the estate area intact with a four-vane fantail but no machinery. The mill provides a cozy veranda where one may entertain friends. The terrace partly circles the base of the windmill. Several mills, whose task of grinding has been taken over by modern technology, have been converted into dwellings.

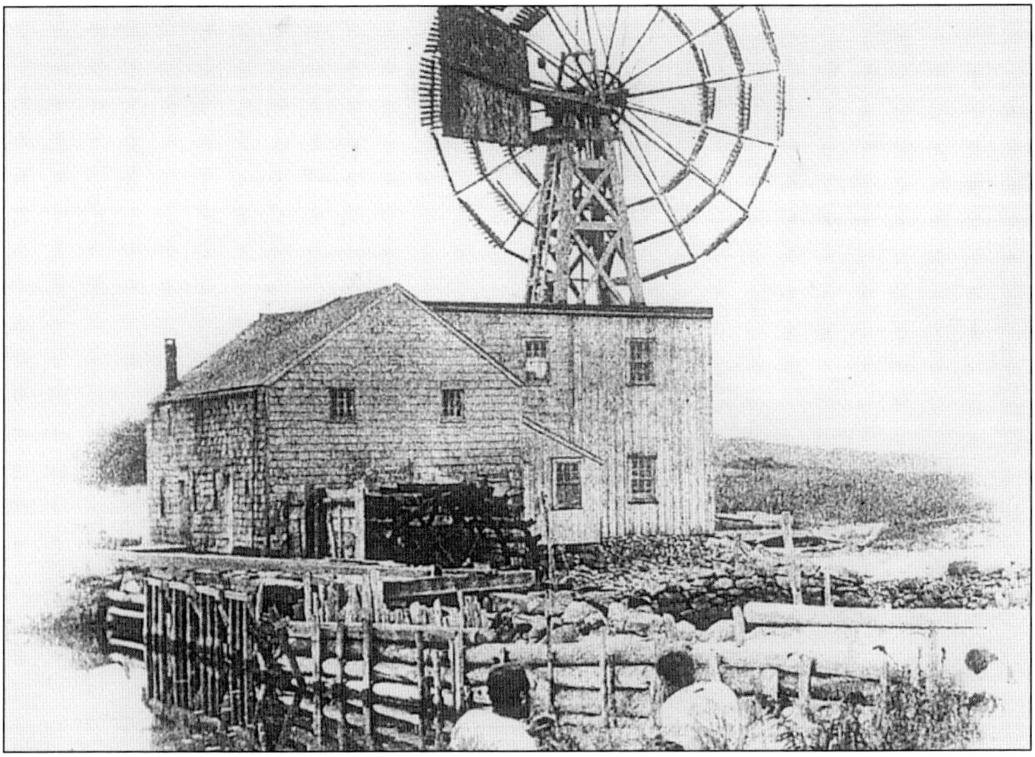

Old Mill in Peconic harnessed the tides of the sea. Water entered the creek through a gate in the dam which is closed at high tide. The head of water was used to turn the mill wheel. It was a conventional tide mill until wind power was summoned to grind when the water was frozen. Old Mill had the capacity to grind 300 bushels of grain daily.

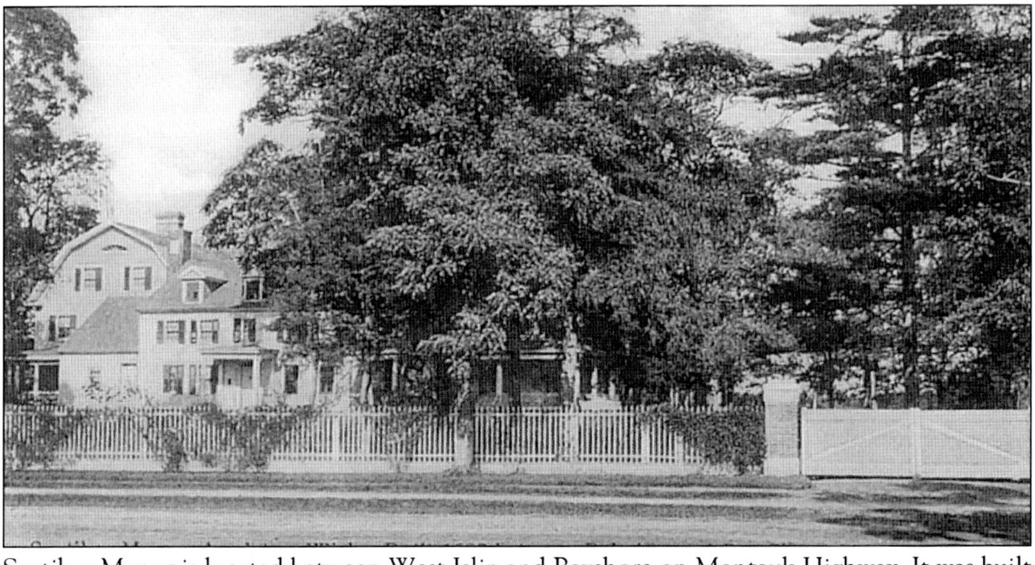

Sagtikos Manor is located between West Islip and Bayshore on Montauk Highway. It was built in 1692 by Stephen Van Courtlandt, a New York businessman. In 1758 Jonathan Thompson acquired the property for his son Isaac, a patriot who was shot by a British soldier. It is now owned by Robert Gardiner and is the headquarters of The Sagtikos Historical Society.

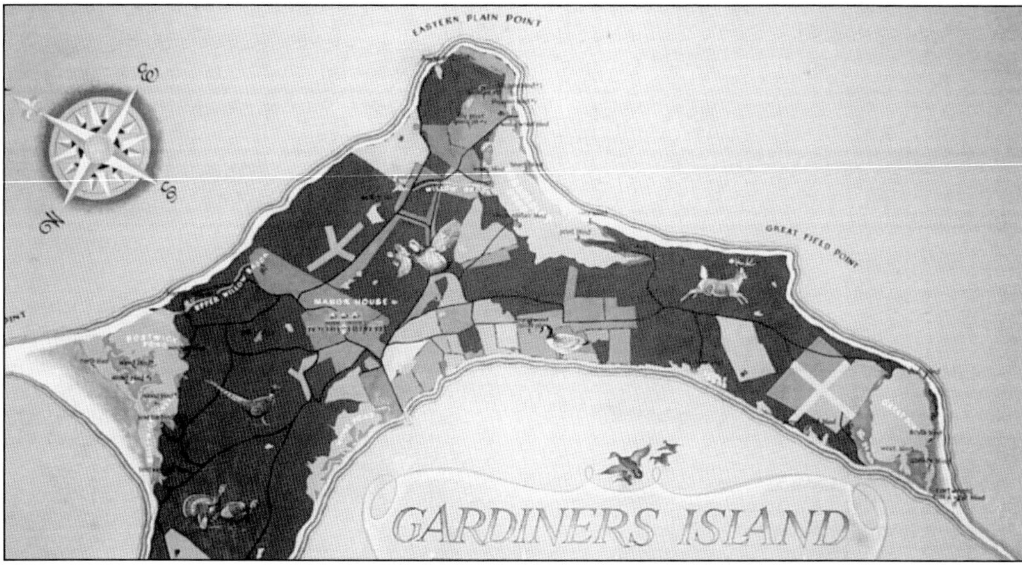

Gardiner's Island lies between the north and south forks of eastern Long Island. It can be reached only by ferry. Lion Gardiner, the first English settler, was a military engineer who bought the island from the Montauk Indians in 1630. Gardiner was born in England in 1599 and came to New York by way of Connecticut. Gardiner Island is the only English Manor that remains in the hands of the original family.

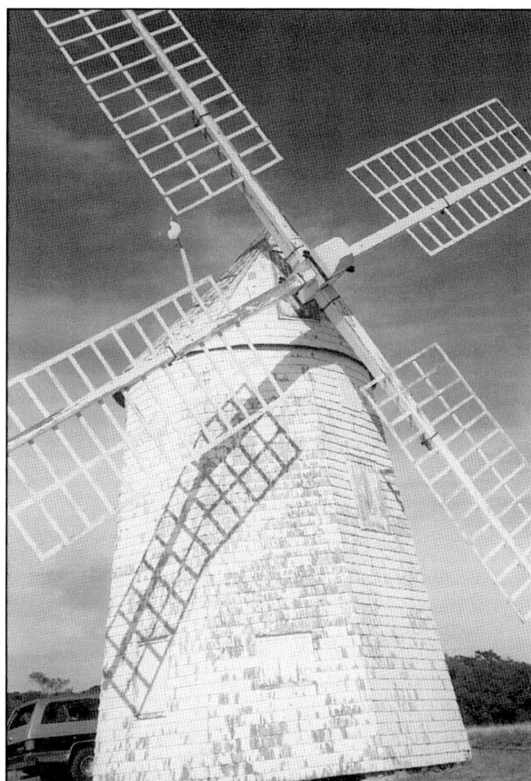

Gardiner's Island Windmill was built in 1771. A boat shaped cap was added in 1815. It is a small mill built with one pair of millstones driven directly from the brake wheel. On May 23, 1795, the mill was moved to an open field. Gardiner hired 34 men to move the mill. A weather vane cock of lead and copper was installed on the mill.

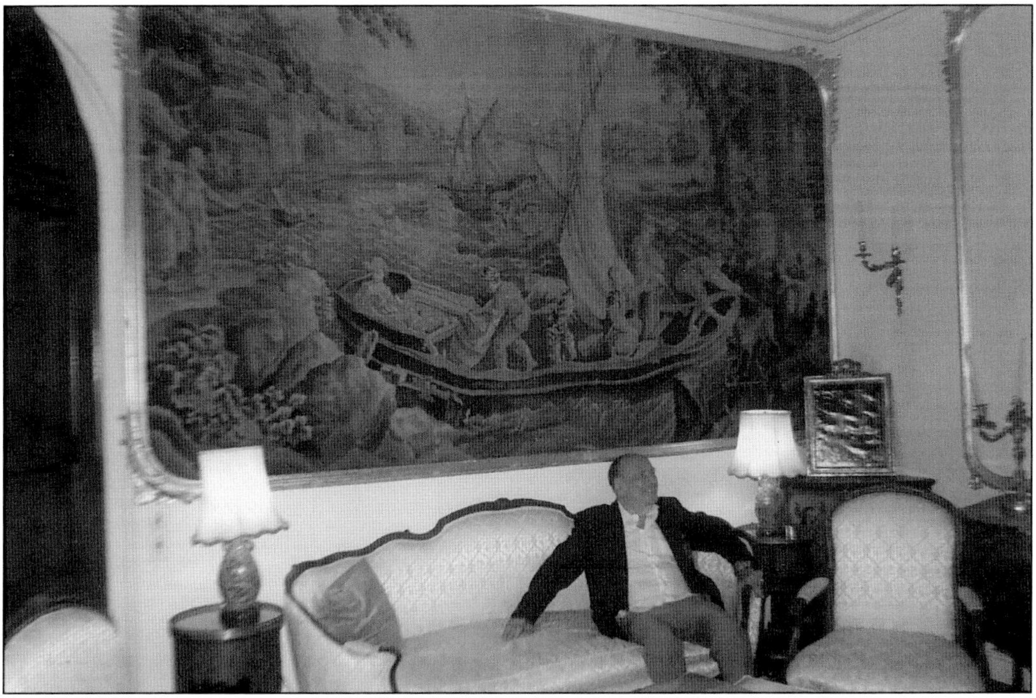

Robert David Lyon Gardiner is a descendant of Lion Gardiner, the original owner. He is the present Lord of the Manor and has been so most of his life. He conducts tours periodically throughout the island. He is pictured in his Easthampton estate beneath a huge depiction of Long Island's earlier days.

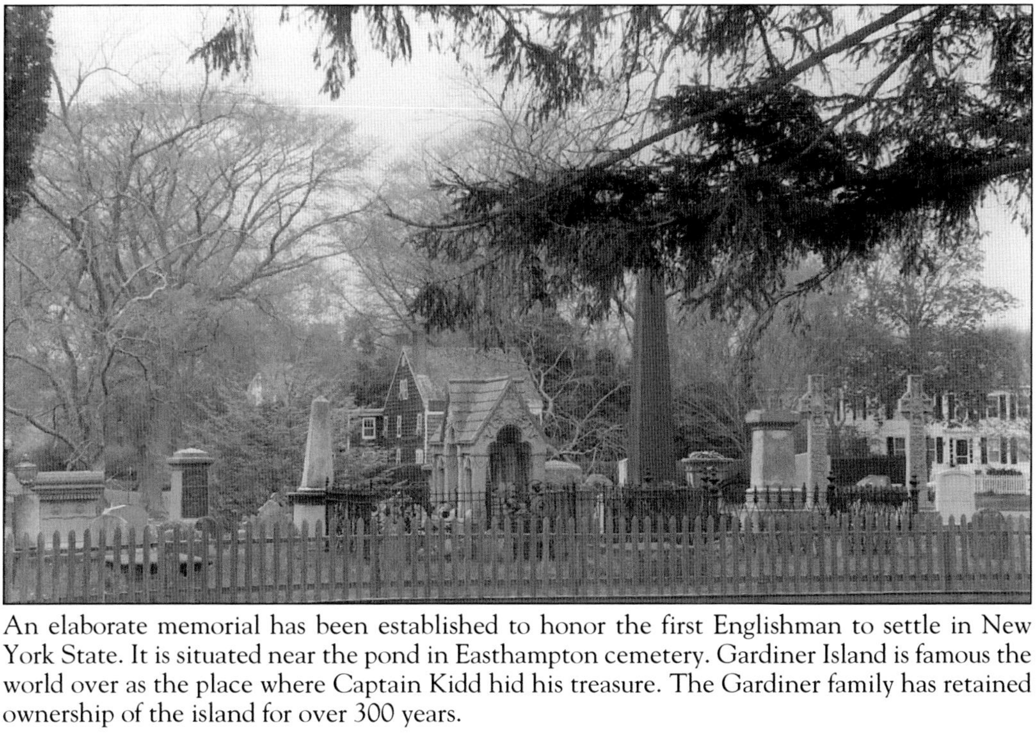

An elaborate memorial has been established to honor the first Englishman to settle in New York State. It is situated near the pond in Easthampton cemetery. Gardiner Island is famous the world over as the place where Captain Kidd hid his treasure. The Gardiner family has retained ownership of the island for over 300 years.

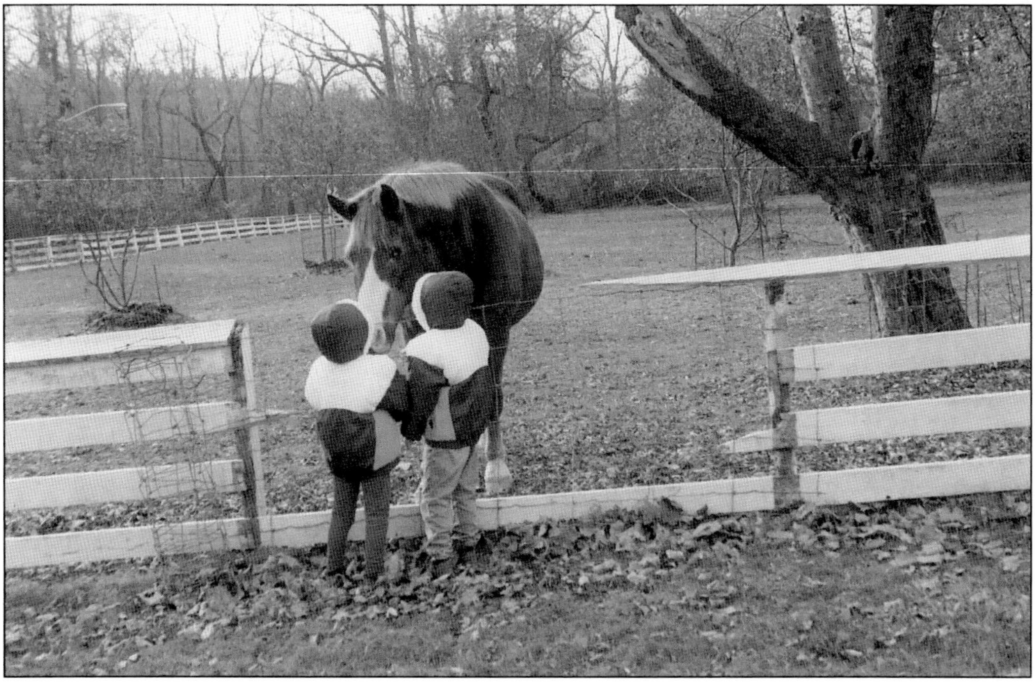

Jamie and Rachel Kriger share some secrets with an understanding friend. He listened attentively and seems to be quite interested in their ideas.

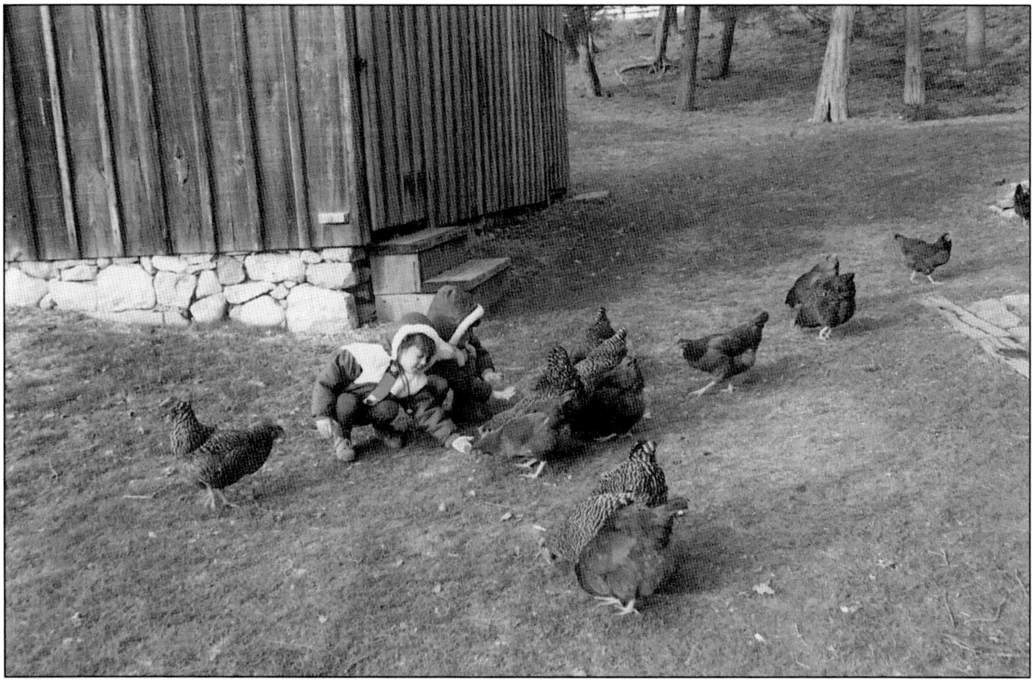

Chickens were a necessary part of country living and were found on almost every Long Island farm. Jamie and Rachel Kriger enjoy sharing a few morsels with the happy brood.

Two

Suffolk's Historic East End

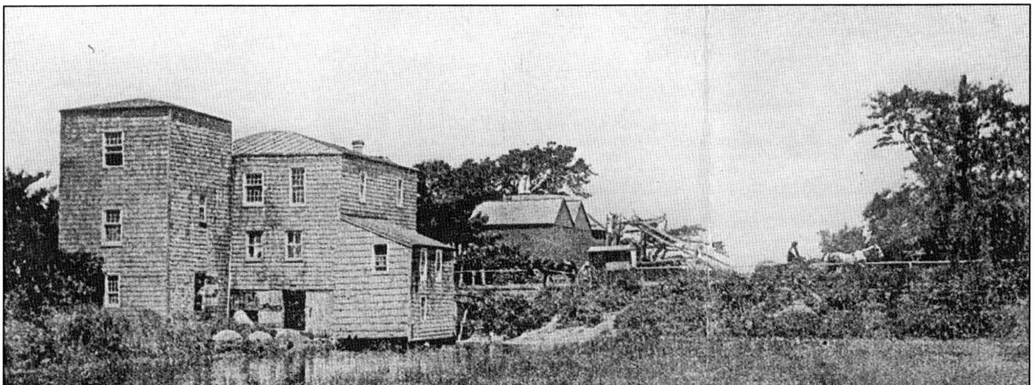

Old Water Mill is the oldest mill on Long Island. Located on Mill Creek in Watermill, it was built in 1644 on the north side of the road. In 1726 the mill was moved south to its present site and gave the village its name. This 1904 photograph shows the waterside of the mill with the Benedict Ice House in the background. Milling operations included grinding grain, fulling wool cloth, spinning yarn, weaving, and manufacturing paper (and, later, ice cream).

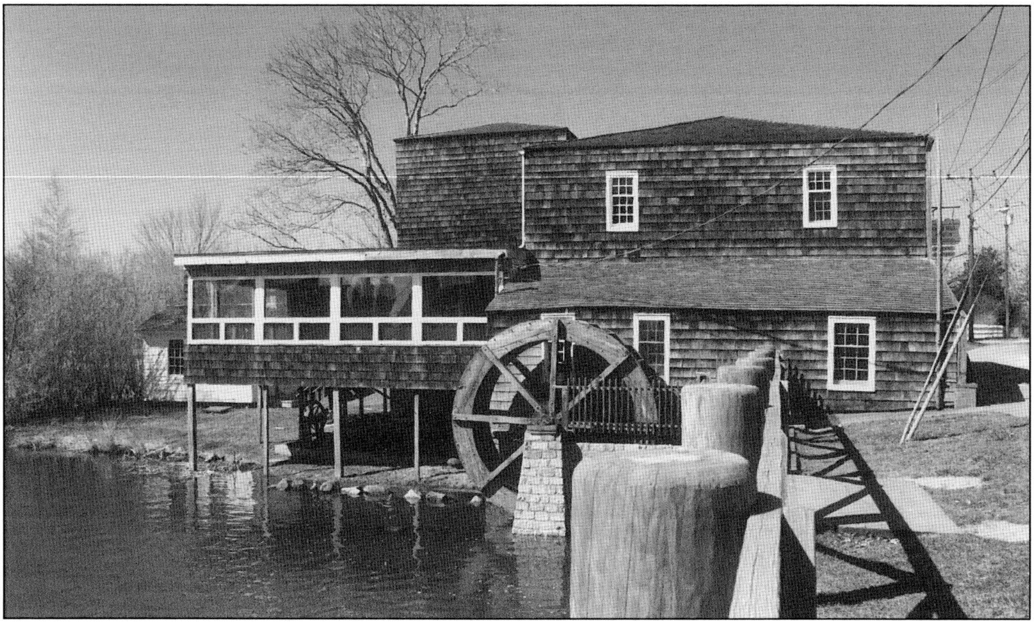

Old Water Mill has become a museum. The public may view many tools related to milling. This includes a display of tools relating to farming, coopers, blacksmiths, weavers, carpenters, ice harvesting and mill work. The museum grounds include Uncle Fred's Workshop. It was here that Fred Benedict plied his many trades and which, today celebrates his creative work with its permanent collection of the implements of another era.

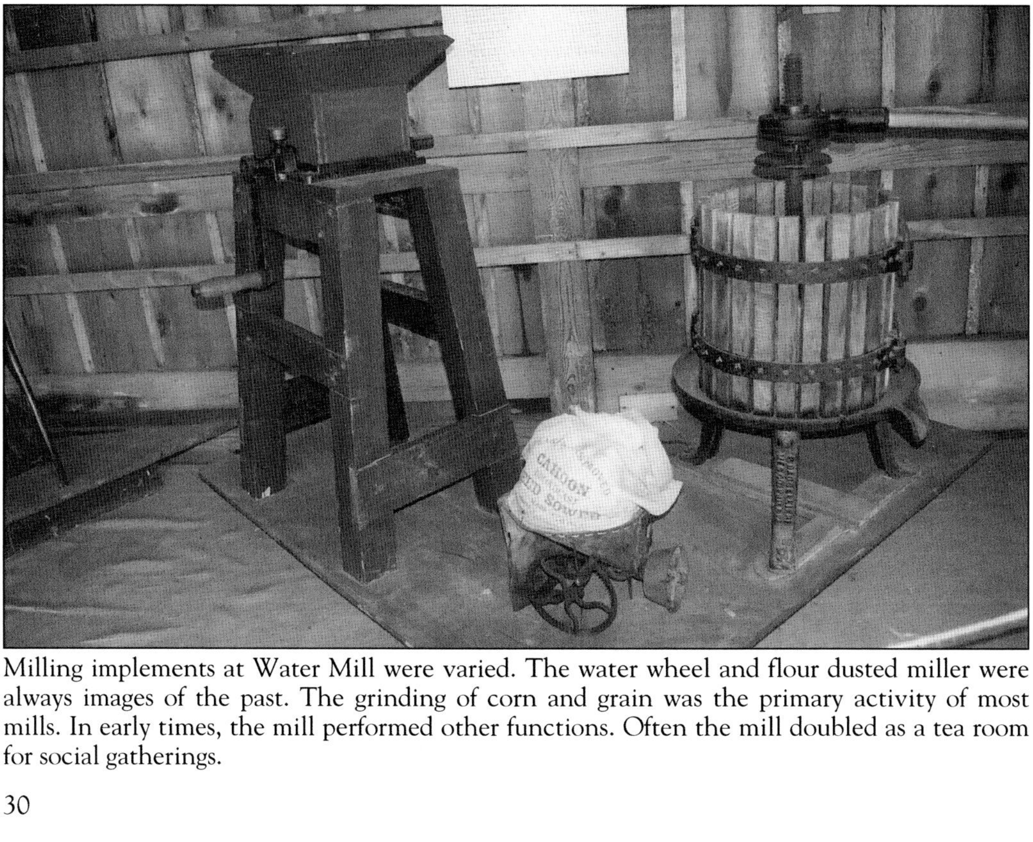

Milling implements at Water Mill were varied. The water wheel and flour dusted miller were always images of the past. The grinding of corn and grain was the primary activity of most mills. In early times, the mill performed other functions. Often the mill doubled as a tea room for social gatherings.

The Ladies Auxiliary of Watermill made the historic quilt of 1987. This organization was founded in 1915 for the purpose of preserving Old Water Mill. In 1968, the auxiliary began to restore the building as a museum. Following years of research, fund-raising, and reconstruction, the historic grist mill was again grinding grain for the bicentennial in 1976. The quilt depicts scenes on both land and sea.

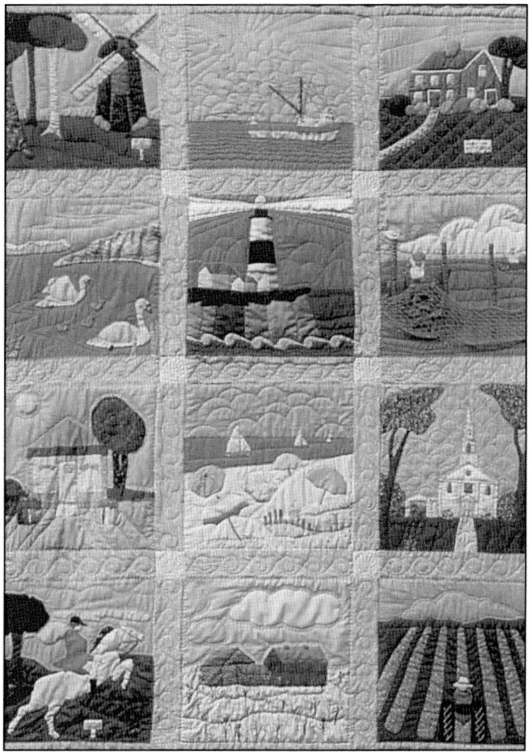

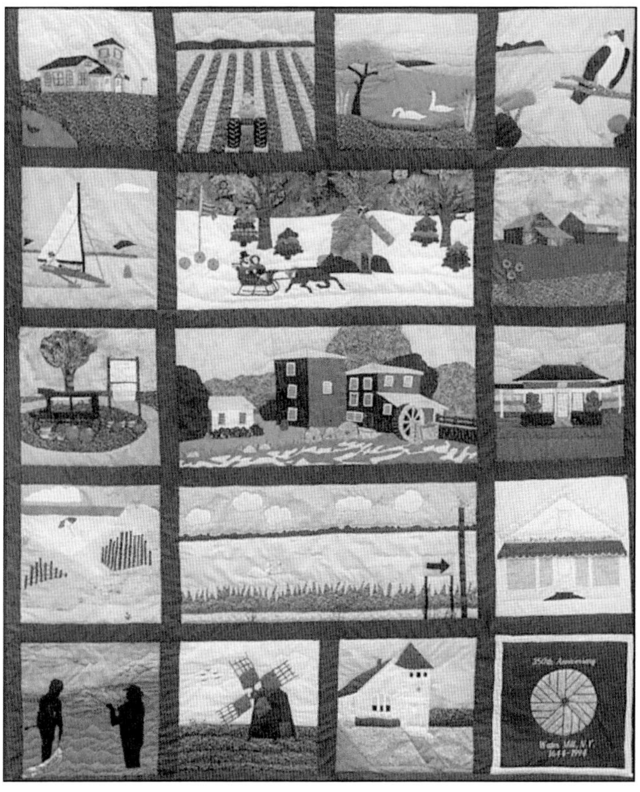

The quilt of 1994 features Old Water Mill. The logo of the Ladies Auxiliary is in the lower right-hand corner. The sailboat, farmland, candy shop, fishing scene, sand dunes, and replica mill on Corwith Farm, with the horse trotting in the snow, were all part of the Water Mill scene.

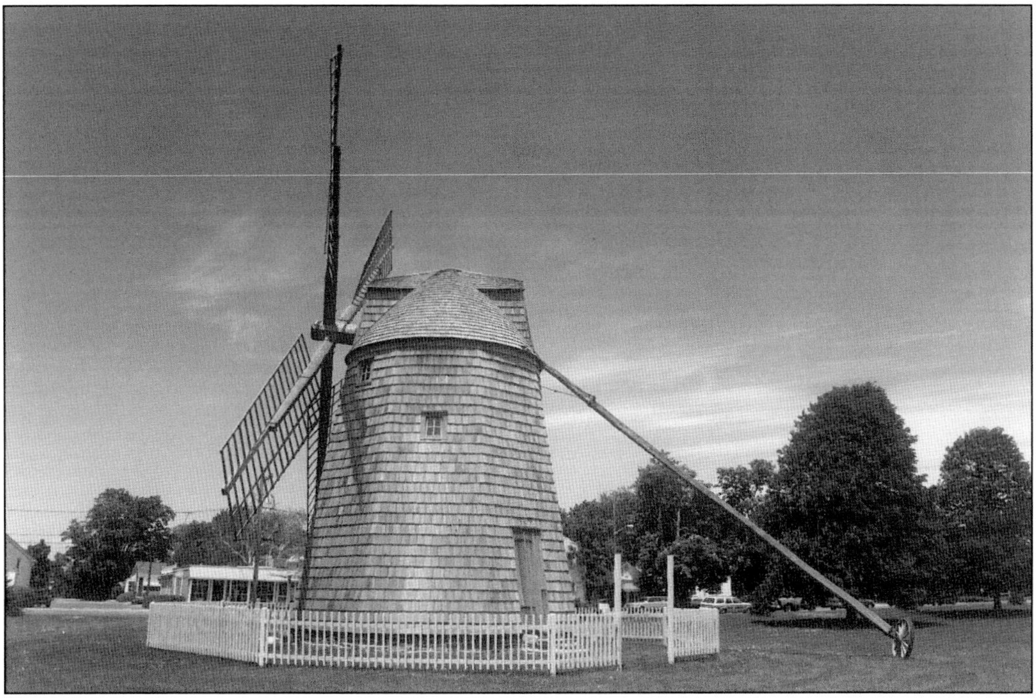

Corwith Mill at Watermill was built on North Haven Neck in 1800. Ox teams moved it to its present location on Montauk Highway by James Corwith in 1814. It operated as a mill until 1887. It has a conical cap with gabled dormer out of which pass the wind shaft and huge tail pole. It features double sided common sails. The wind shaft is unusual with neck and tail of cast iron and a wooden pole on the front of the neck for holding the sail stocks.

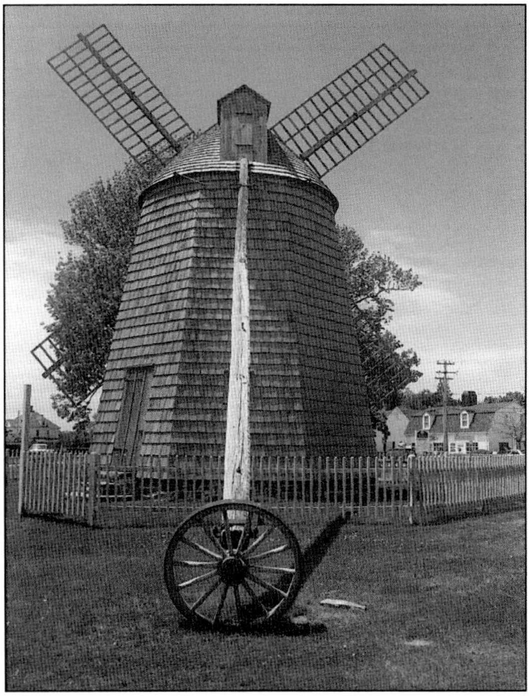

A tail pole or sweep turned the Water Mill cap. This was a long piece of timber extending from the rear of the cap to the ground. The tail pole provided greater stability in severe winds. This part of post mill technology was still valued in mill construction in exposed locations. It is a carry over from the post mill.

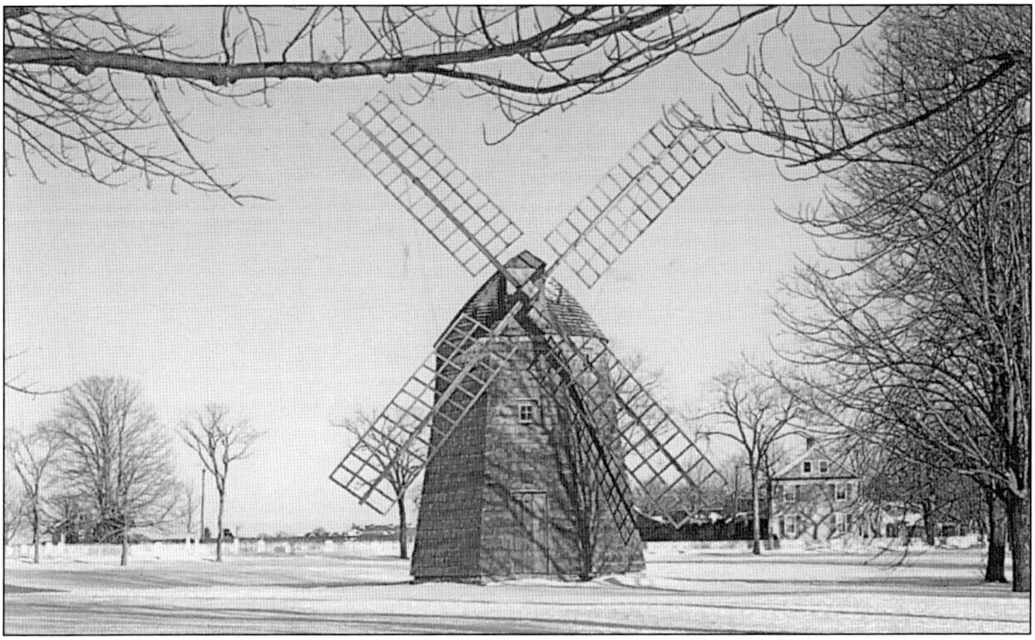

Water Mill was among the traveling mills. Mills were constructed of wood and were easily dismantled. James Corwith bought the mill from J. Mitchell in Hogs Neck for $750.00 in 1813. Twelve yokes of oxen transported it to common land in Watermill. Fall was the traditional time to move and reconstruct mills. A second set of millstones was added and the original system replaced. A board inside the mill is carved "Began to Grind August 1, 1800." It operated successfully for the next 75 years.

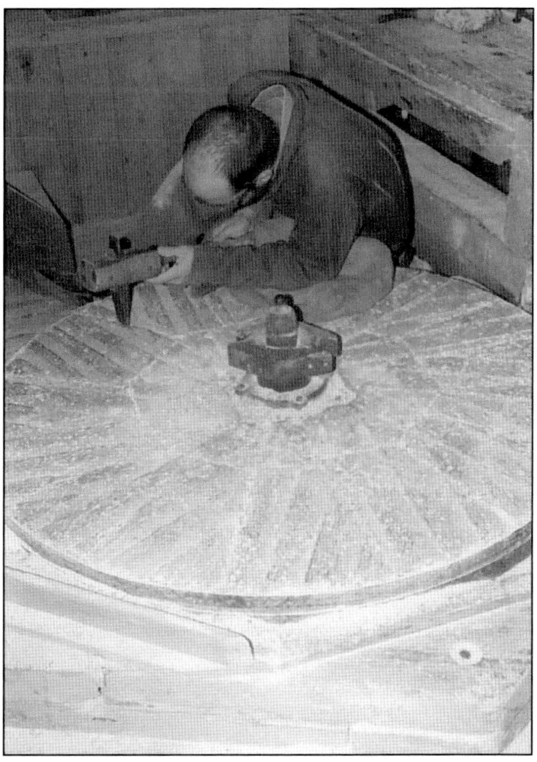

A miller is dressing the bed stone. He prepares a smooth surface by making it slightly hallowed toward the center. Using a special tool, patterns of furrows are carved into the stone. They are arranged so that when the stones are in position the furrows crossed at each revolution. This was done with precision and skill. This is what performs the actual operation of grinding.

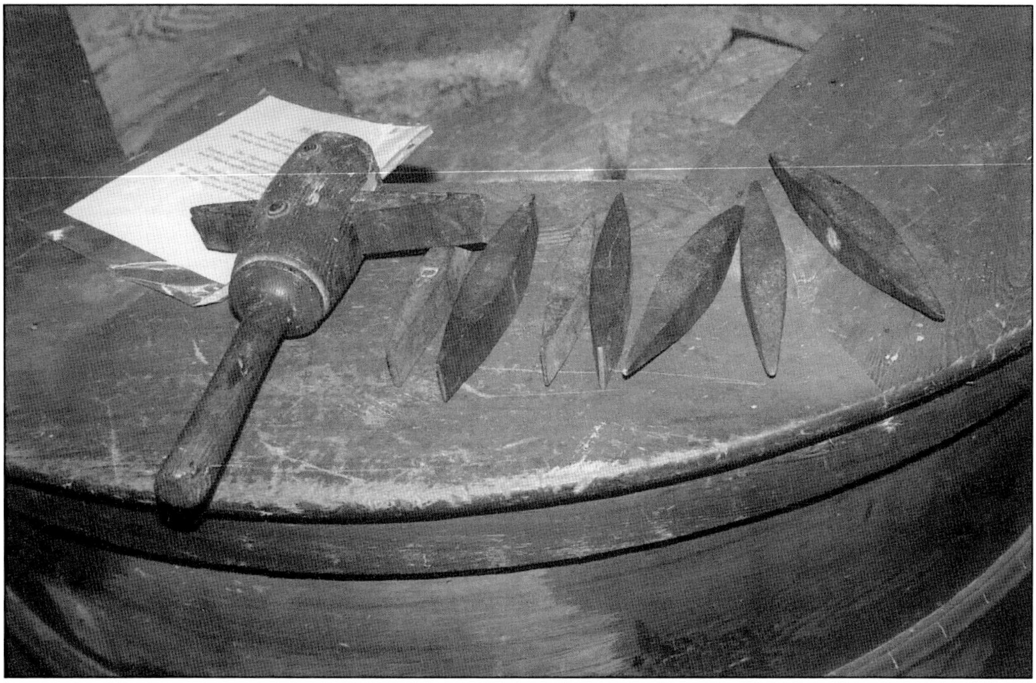

Devices for dressing a bed stone were necessary because constant use dulled the stone. The cutting tools were picks and bills used by hand. The runner or capstone had to be removed. This was done with the help of a hoist or crane. The actual carving of furrows was done by hand. The capstone always had a hole in the middle that fit into the bed stone.

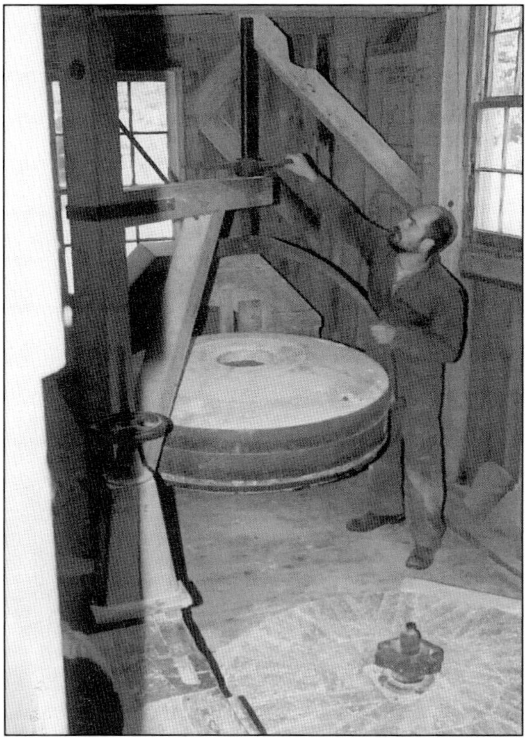

The miller is adjusting the capstone. It has been removed from the bed stone and will be placed up side down on the floor until dressing is complete. Tendering the stones is a process of altering the distance between the stones in order to maintain a uniform degree of fineness. The two millstones are never permitted to touch The experienced miller can detect the smell of granite indicating the stones are too close. He then adjusts the gap between the stones.

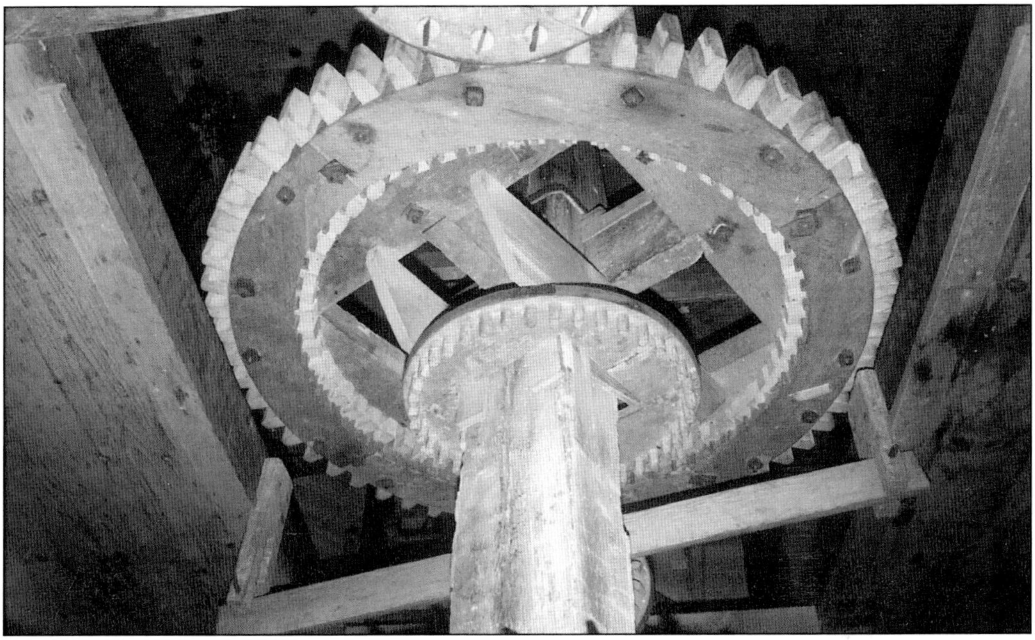

The machinery in any mill is intricate. In Corwith Mill the drive for the machinery begins on the upright shaft with wooden teeth facing upward. The drives and wooden pinions are at right angles on a horizontal shaft. The shaft is mounted on a pivoted piece that can be engaged at will. Many mills have weather vanes in the gables.

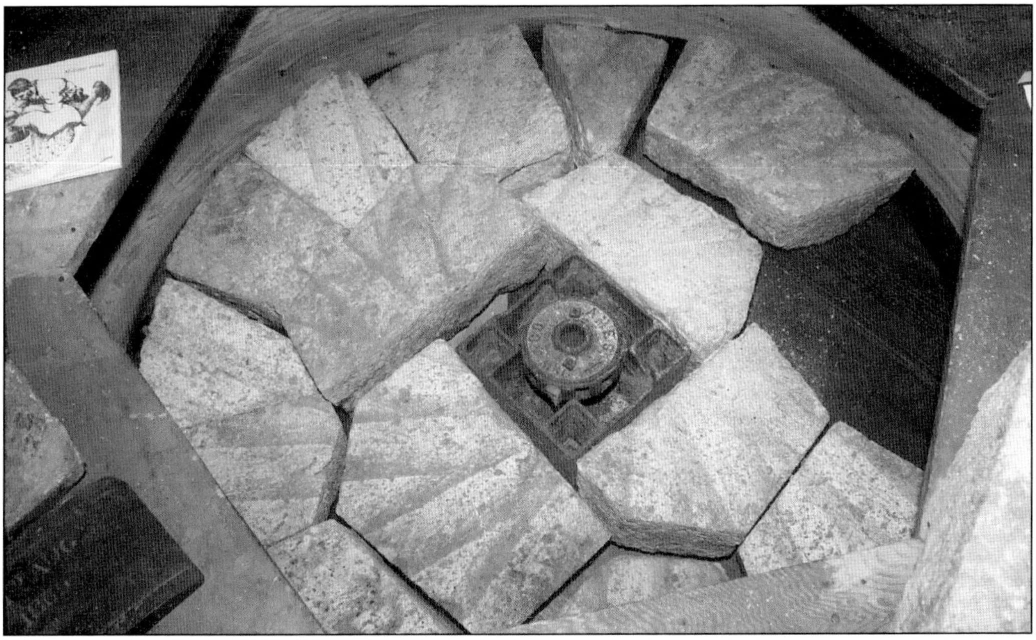

Burr Stones are high quality French quartz millstones. The small blocks, quarried near Paris, were shaped and matched together into a rounded stone and cemented together. They were bound with iron hoops and backed with plaster of Paris. Two sets of stones operated in the Corwith mill and one was the burr stone. Burr stones are harder than those from American quarries and they require less frequent dressing.

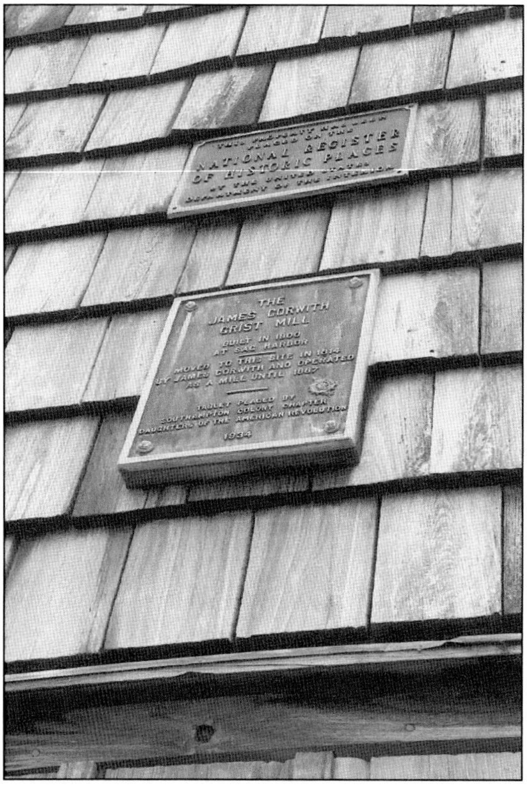

Corwith Mill at Watermill was designated an historic site. The plaque reads: "This property has been placed on the National Register of historic places by the United States Department of the Interior. The James Corwith grist mill built in 1800 at Sag Harbor was moved to this site in 1814 by James Corwith and operated as a mill until 1887. The tablet was placed by Southampton Colony Chapter, Daughters of the American Revolution in 1934."

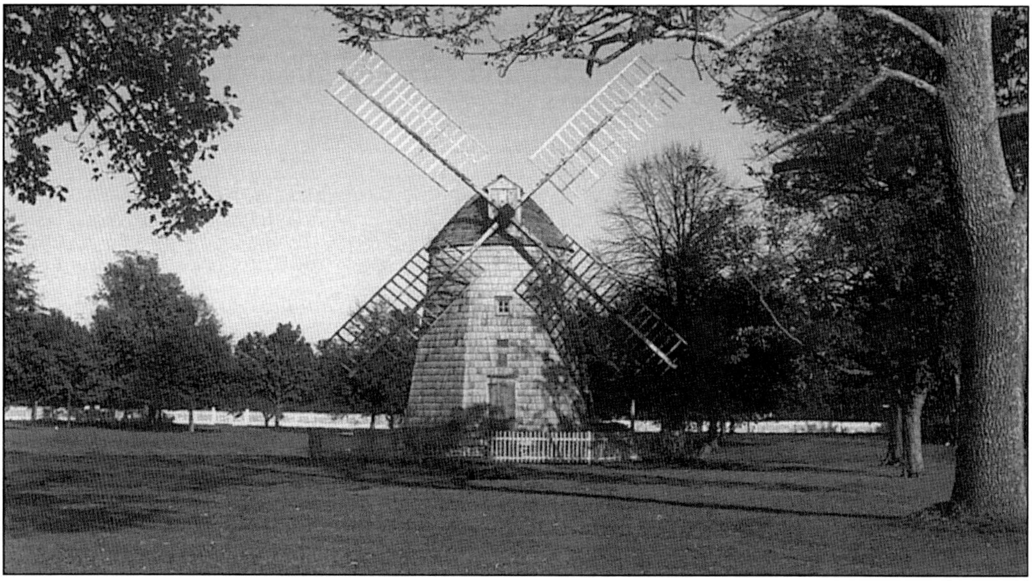

Corwith's mill has undergone restoration in the last decade. The citizens of Watermill undertook a major restoration project in 1984. They wanted to make the grinding, whistling, clunking, singing, and merry groaning of the windmill come alive for the children of the millennium and beyond. The experiences of many generations are now accessible at Watermill. For many years, the mill had been a gathering place for children during Santa's annual visit to Watermill.

This replica windmill was erected on James Corwith's Farm. These wonders of another age symbolize clean, efficient, cost-free technology. They need no fuel and cause no pollution. The windmill uses a free resource to pump water, and generate electricity. This little windmill is a reminder of an ecologically perfect machine.

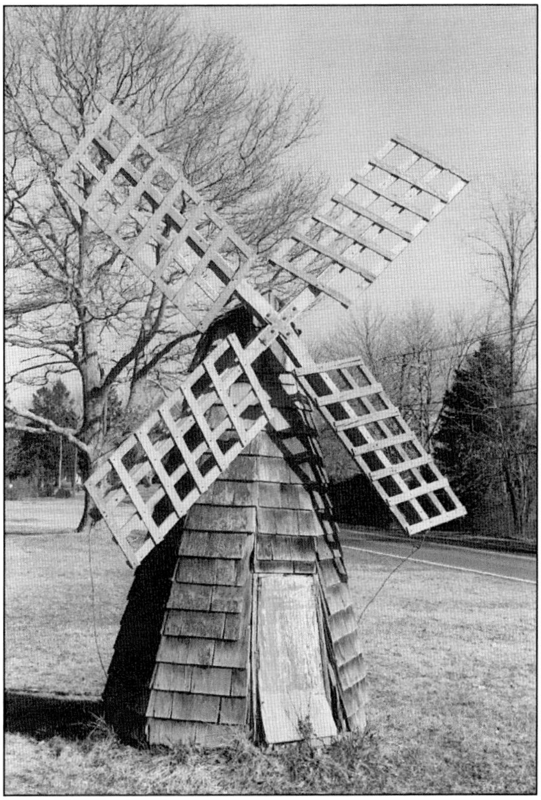

Caroline Church of Brookhaven in Setauket was built in 1729. It was named in honor of Caroline, Consort of King George II. The queen sent a silver communion service with embroidered alter cloths as a gift. It was lost during the revolution. The church was used as a hospital during the ""Battle of Setauket." The building still displays the bullet scars of 1777. It is a fine example of colonial church architecture. It is the oldest Episcopal edifice on Long Island.

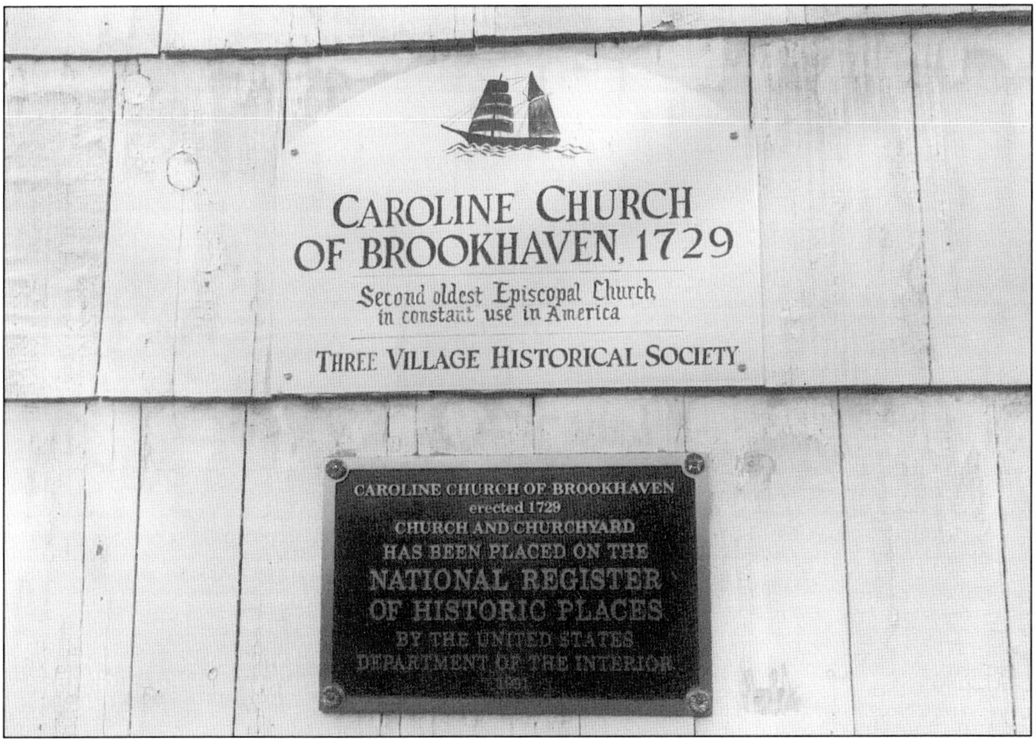

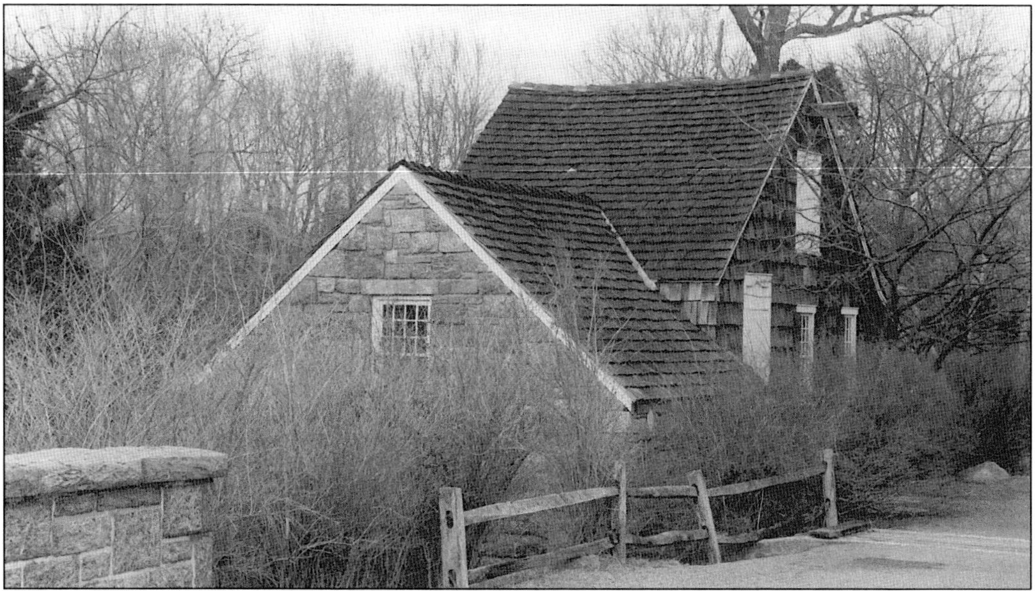

The Village of Setauket, on Conscience Bay, has had several mills dating back to 1664. John Wade bought the third one and became a miller. He was followed by Richard Woodhill and Issac Satterly. The last miller was Everett Augustus Hawkins who closed the mill in 1930. The present mill is simulated and commemorates the old Setauket Mills. It emphasizes the necessity of mills in the 17th and 18th centuries.

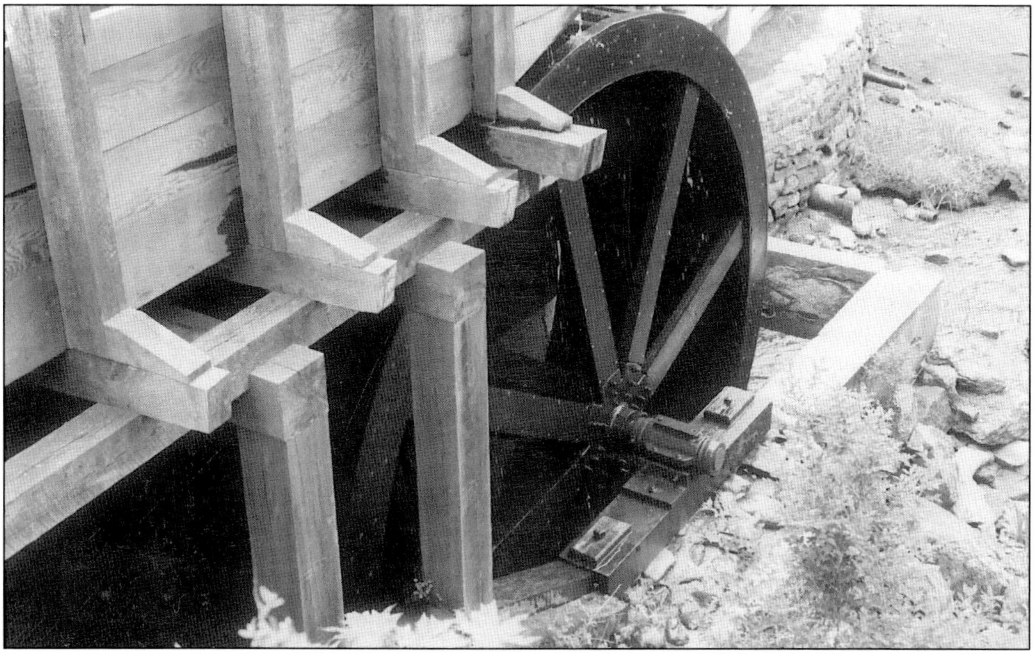

The water wheel of the original mill has been installed into this replica water mill. It was constructed in Setauket in 1956. Located behind the post office it was built by Ward Melville, a local promoter of historic preservation. The need for mills began to decline in the 1880s. Newer techniques were initiated, and then the railroad connected Long Island farms to New York City.

This replica farm windmill at Holtsville was built on the Holtsville Ecology Site. Its predecessor was a farm mill southwest of Gibbs Pond Road. Axioms have stemmed from the milling business. The miller was a reliable judge of the quality of grain. He often tested its coarseness and moistness by rubbing a pinch between his thumb and forefinger. From that action came the term "Rule of Thumb."

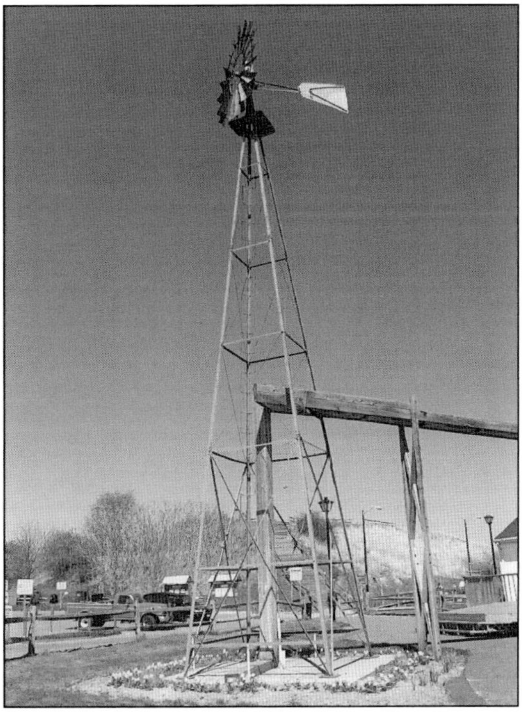

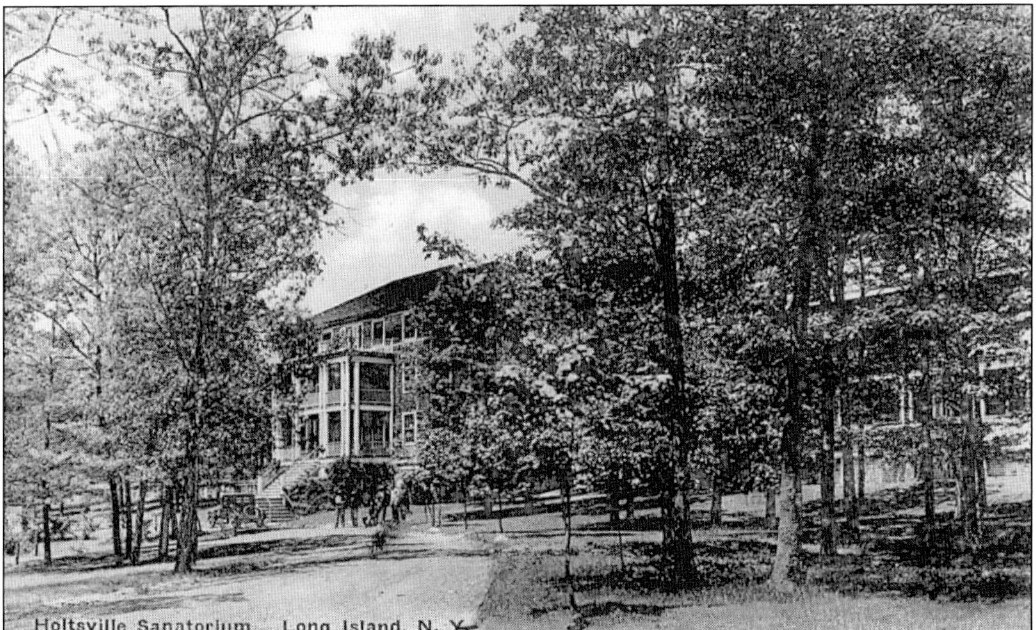

Suffolk County Sanitorium was built in 1916, 250 feet above sea level, in Holtsville. The facility accommodated over 100 patients. In 1960, the "San" was discontinued. Suffolk Community College was about to open. The administration building was located here until the Selden Campus was ready. The opening enrollment in September 1960 at Holtsville was 506 students. In 1961, 1,440 students enrolled the first semester on the new 170-acre Selden Campus.

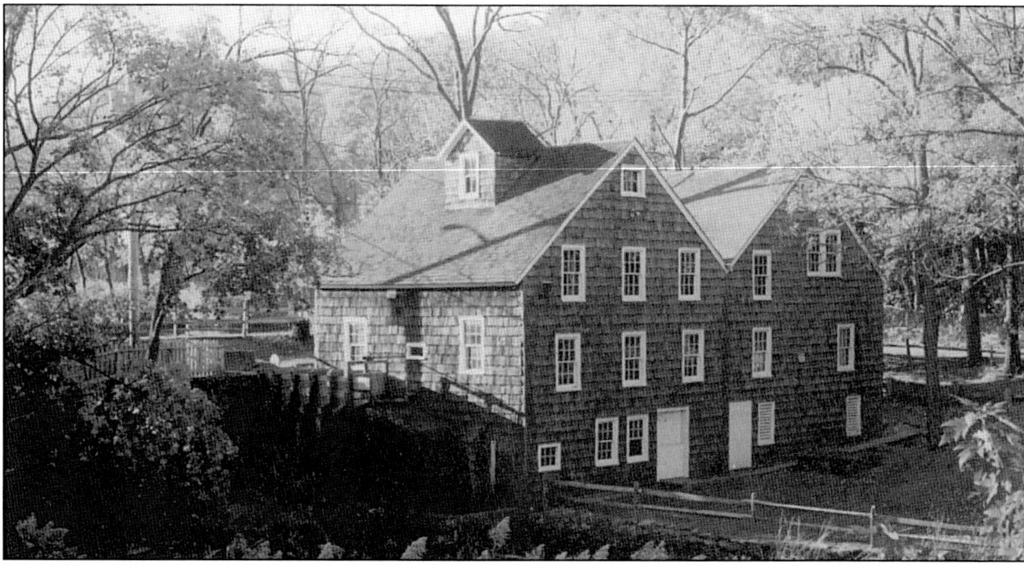

Stony Brook Grist Mill dates back to 1698 when Adam Smith obtained water rights to the stream. The original mill was destroyed and washed away when the dam broke in 1751. Plans were immediately in place for a new mill. In 1751, the present mill was erected on the Smithtown side. The Smith family operated the mill for many generations. In 1776, British troops obtained supplies here. Stony Brook is the last of water powered grist mills to operate on Long Island.

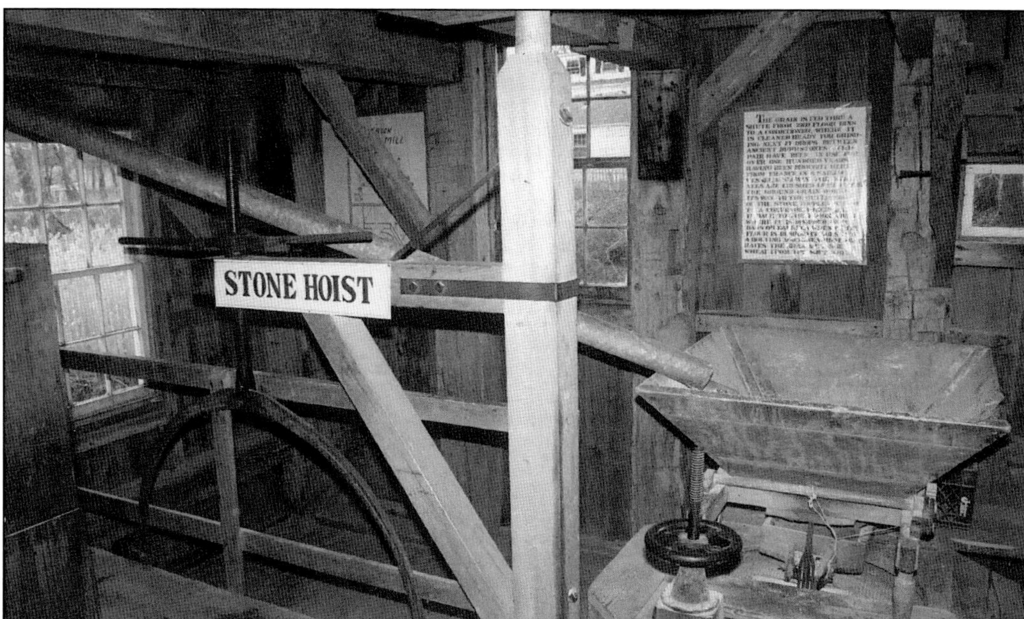

The stone hoist at Stony Brook Grist Mill resembles old ice prongs. They were used when the iceman delivered a huge block of ice and placed it in the icebox. The hoist in the mill is used to lift the top millstone, the capstone. Using the stone crane, the miller can lift single-handedly a one-ton millstone. Mills not only ground grain and corn but also sawed logs. Mills, blacksmiths, and shipyards were the most respected proprietors on Colonial Long Island.

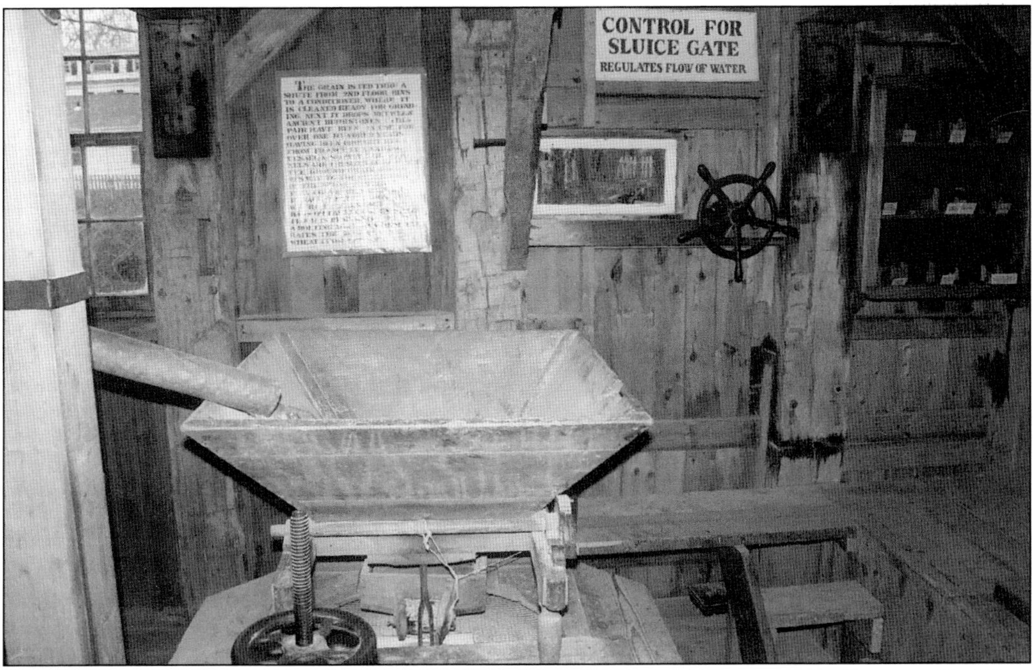

The interior of the Stony Brook Grist Mill has something in common with all gristmills. They have a language all their own. The sluice gate is a device that holds the water back and regulates its flow. It may also be considered a floodgate. The sluice is a wooden trough that carries water to the water wheel.

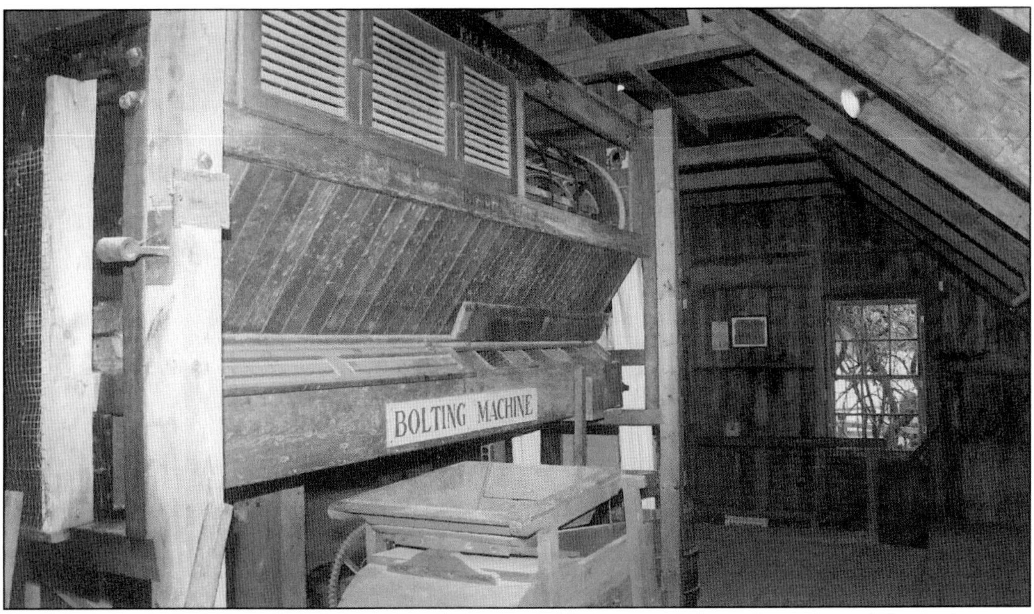

The bolting machine in the Stony Brook Grist Mill separates flour from bran. The bolter is a square framework of wooden bars held by spokes to an axle. Silk was nailed to the bars and the reel revolved in a pine casing which had a trough underneath. The meal in the bolter tumbled about causing fine flour to pass through the mesh of silk and the bran to remain. The bolter can be regulated to sift fine as well as coarse flour.

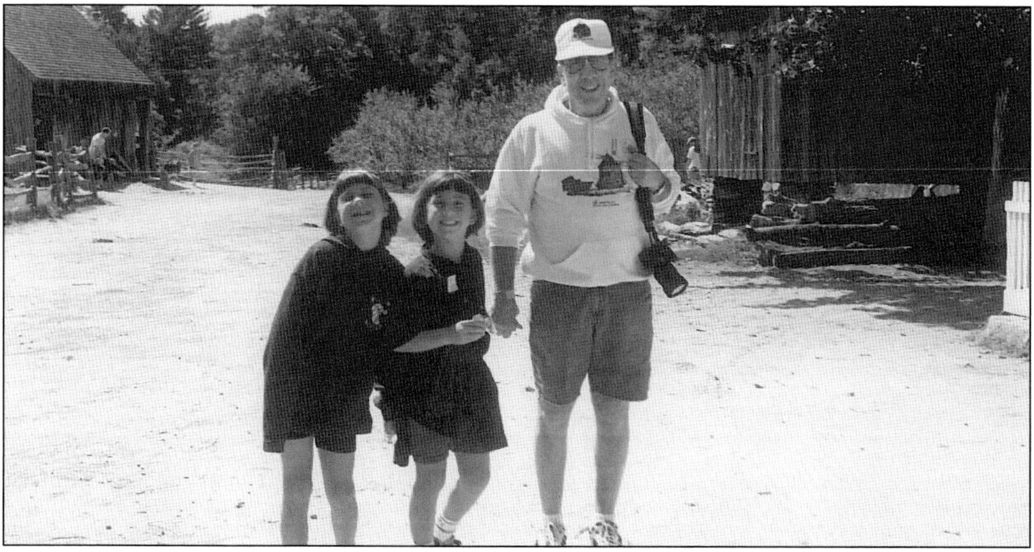

Miller Gerald Leeds and his twin granddaughters, Jamie and Rachel Kriger operate the Stony Brook Grist Mill. Children act as dusties. They husk corn, separate kernels, bag cracked corn and learn the art of milling through hands on experience. The Stony Brook Grist Mill has become part of the expansive Suffolk Museums. A huge carriage house containing wagons of a bygone era, restored houses, a steam locomotive, blacksmith shop and colonial village are just a few of the mementos of yesteryear at Stony Brook.

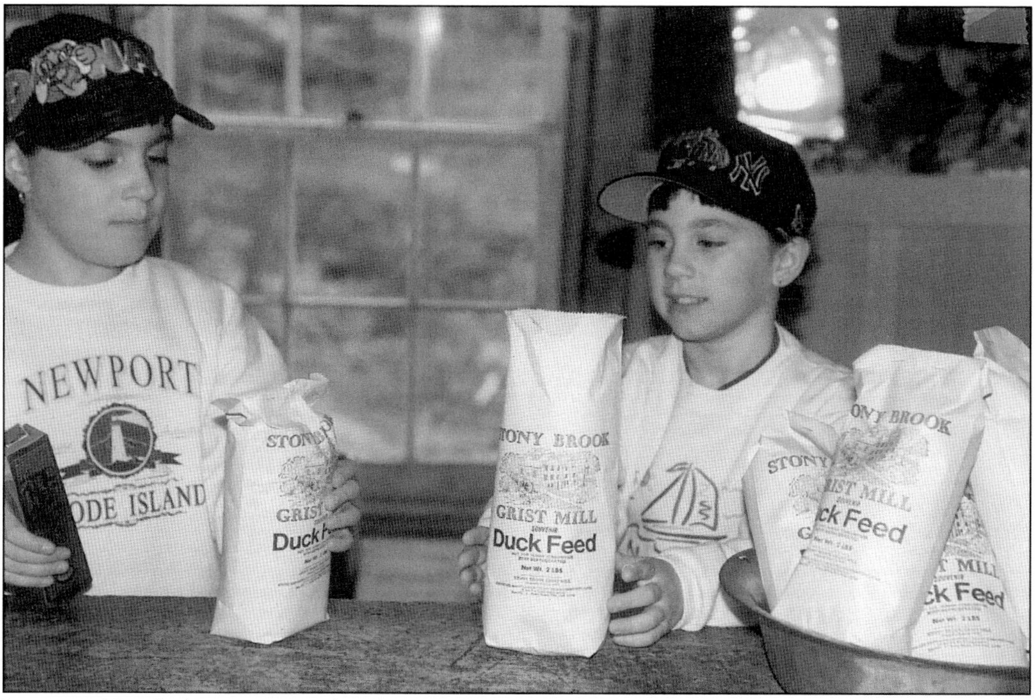

Jamie and Rachel Kriger, twin daughters of Steven and Lisa Leeds Kriger assist their grandfather, Miller Gerald Leeds, at Stony Brook Grist Mill. They are bagging duck feed for the many duck farms on Eastern Long Island. The yellow sacks are designated as souvenir bags. The girls are familiar with the workings of the gristmill.

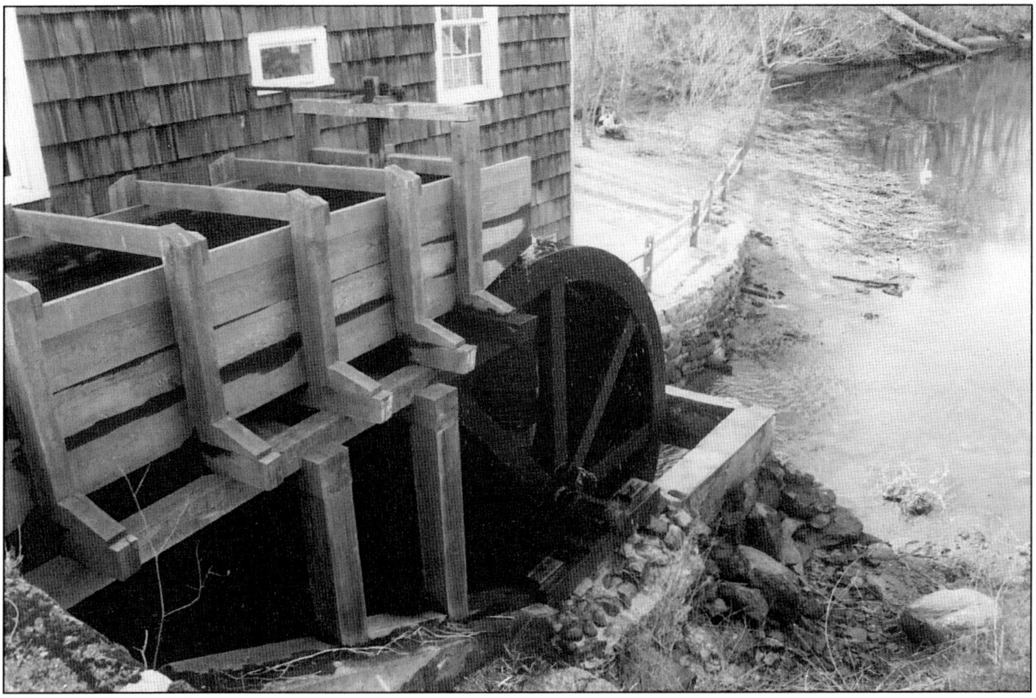

The waterwheel at Stony Brook is known as an overshot waterwheel. When water is dropped on the wheel from above it fills the buckets. The weight of the water in the buckets causes the wheel to turn.

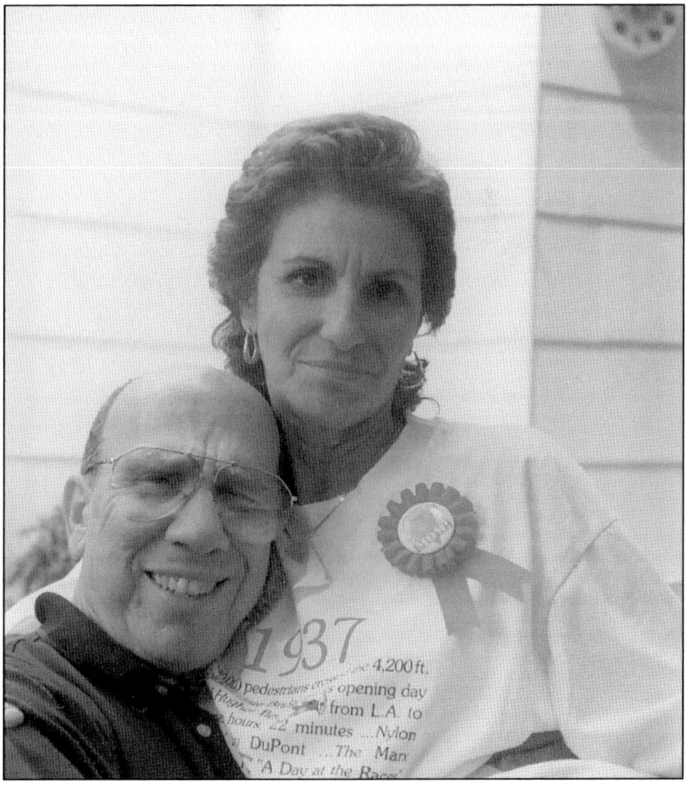

Gerald Leeds (Jerry) is co-director of the Leeds Corporation of Long Island with his wife, Barbara. A retired Health and Physical Education teacher, he presents slide shows on landmarks of historical Long Island, and does in-service educational programs and workshops for teachers. He is also a member of the Society of the Preservation of Old Mills (SPOOM) and The International Molinological Society (TIMS).

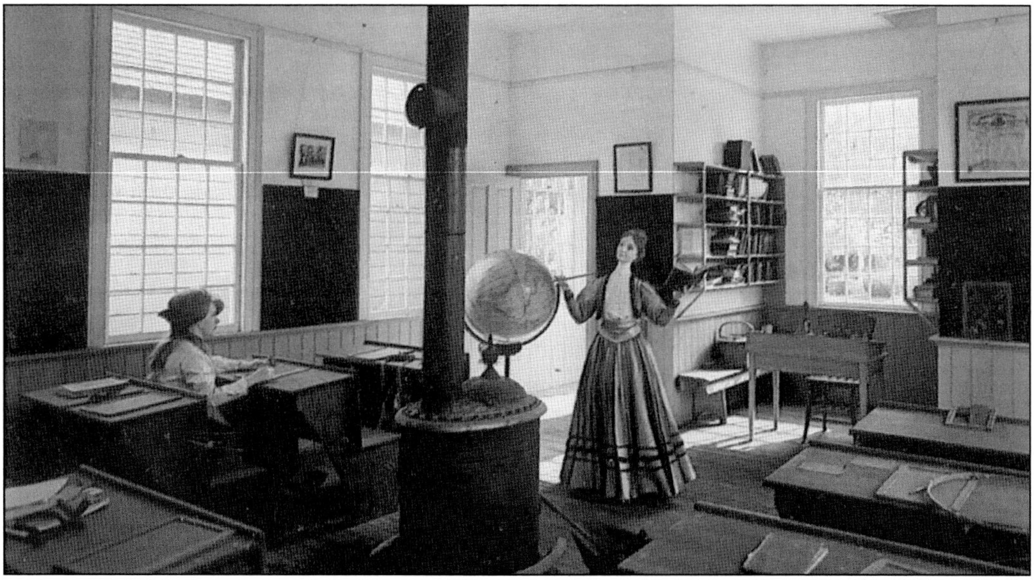

The Nessakeag Schoolhouse was built in 1818 and has been relocated to Stony Brook. It was in operation for over a century. There were separate entrances for boys and girls. The stove supplied heat and slates were the order of the day. It was moved to the grounds of the Suffolk Museum Carriage House at Stony Brook in 1955.

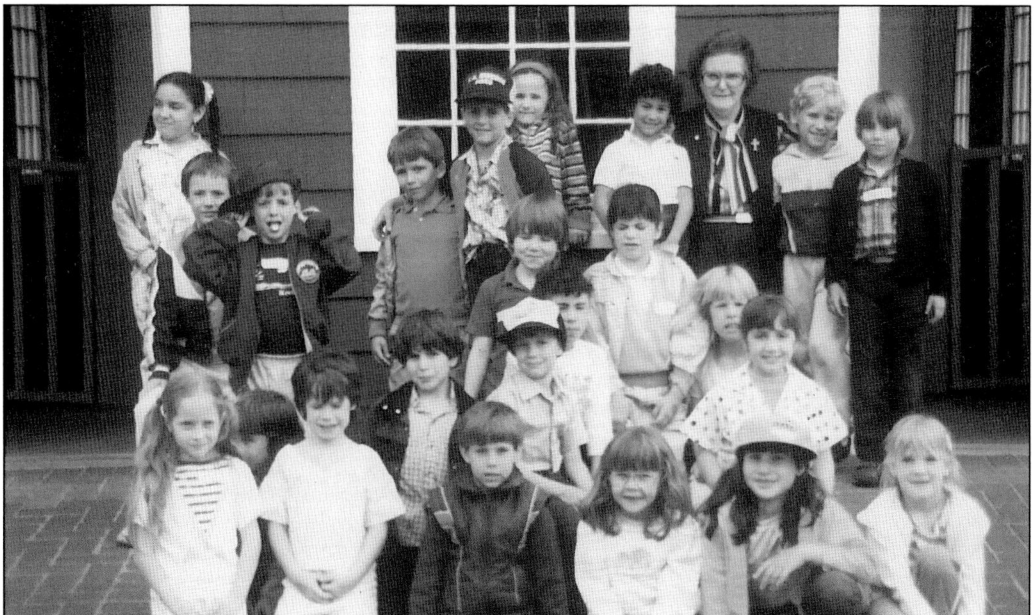

School groups visit Stony Brook frequently. This group from St. Joseph School in Babylon is enjoying the old school house with their teacher, Sr. Anne Frances. Gathered at the entrance are Edward Bradley, Edward Bryne, Andrea Conley, Brendan Cuff, Jennifer Dante, Kevin Davis, Tina Hiltl, Jacqueline Kraker, Karen Lingenfelter, Anne Martin, Cecilia Mejia, Timothy Miller, Melissa Murphy, Kerri Ohman, Paul Neil, Ryan Peinkowski, Jason Pulver, Christopher Sloan, Kevin Smith, Michelle Stein, Lorraine Strawsacker, Karen Tanner, Jennifer Walsh, and Marissa Woodland.

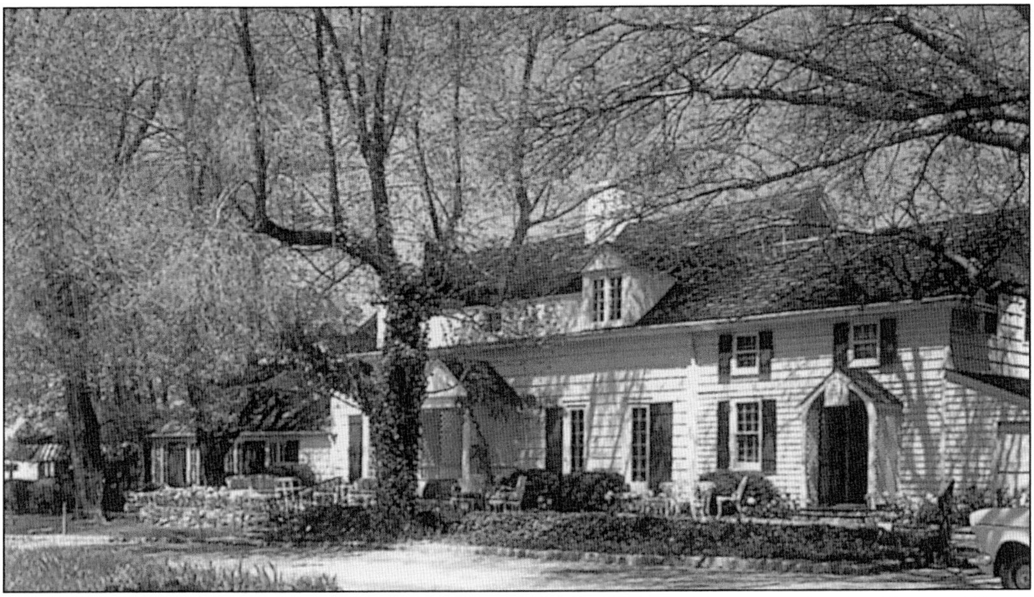

Three Village Inn was built in 1751 for Captain Jonas Smith. He was a shipbuilder and Long Island's first millionaire. Four old fireplaces burn constantly during the winter months to create a homey-long ago atmosphere. In 1930, Ward Melville's mother used the house for meetings with local women of the village. They began serving tea and thus began a tradition that evolved into today's charming restaurant and lodging facility.

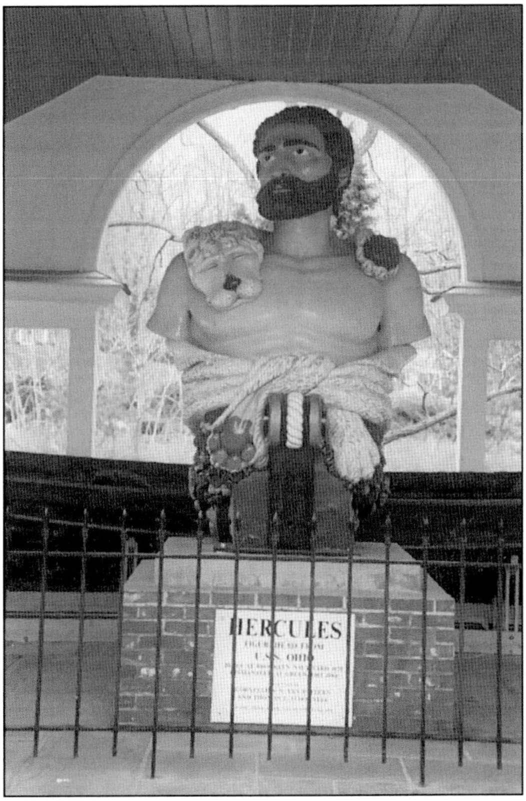

Hercules is a wooden figurehead from the USS *Ohio*, launched in 1820. Following an active career the vessel was dynamited near Greenport and Hercules was brought to Canoe Place Inn at Hampton Bays where it stood for six decades. In 1954, Hercules was brought to Stony Brook where he overlooks the Village Green. Hercules and the anchor beside him were first launched in the Brooklyn Navy Yard in 1820.

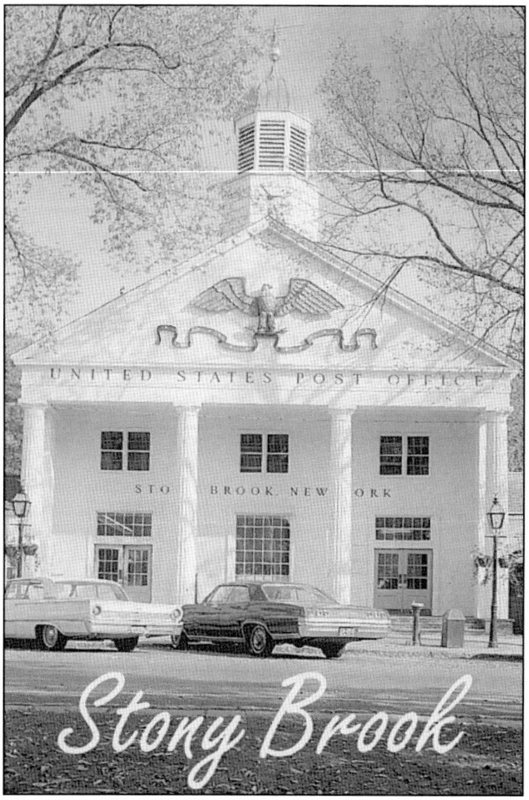

Stony Brook Post Office is the focal point of the colonial shopping center. The American eagle is on its facade, above the Doric columns. It is carved of wood and painted appropriately. People come from miles around to see its wings flap when the clock above, in the cupola, strikes the hour.

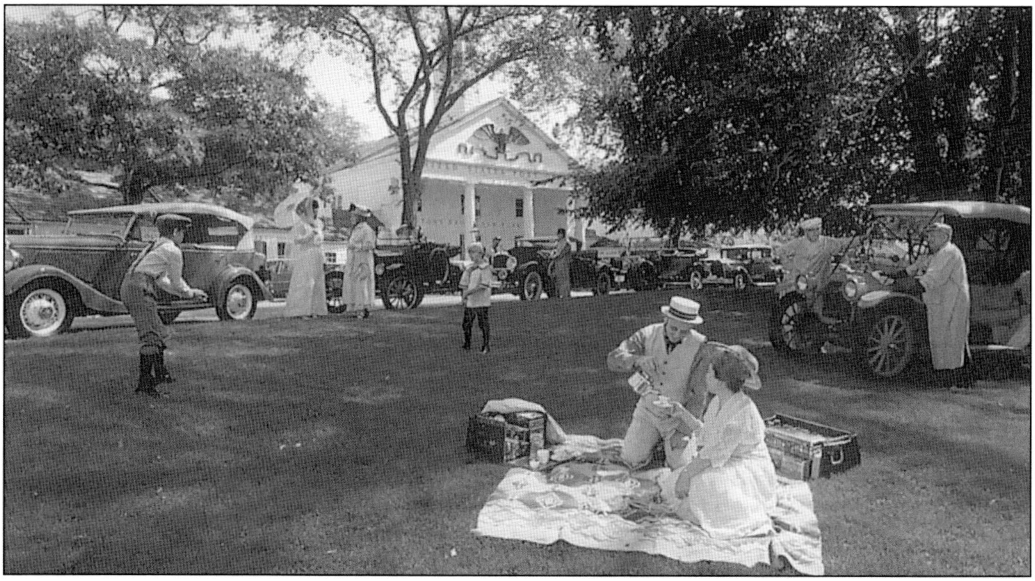

Village Green at Stony Brook is an ideal place for picnics. Here one faces the post office and Main Street. Many of these shops are leading national retailers such as Jones of New York, Laura Ashley, the Nature Company, and Chico Casual Clothing. Created in 1940 by Ward Melville the two-acre Village Green and shops are owned and maintained by The Stony Brook Community Fund.

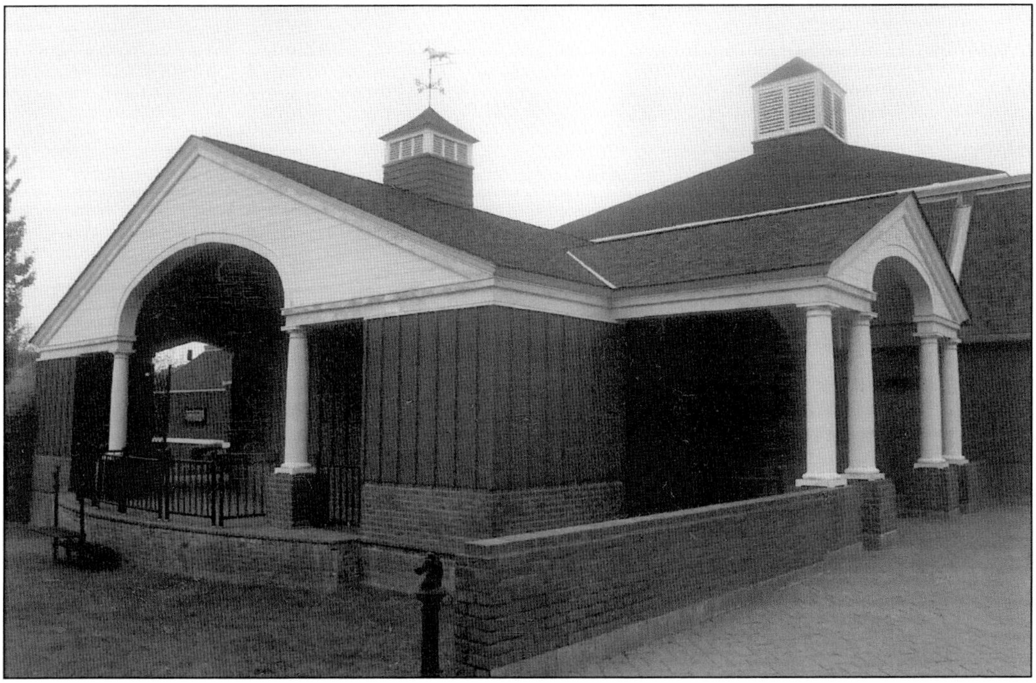

The fame of Stony Brook originated in 1940. Ward Melville, a New York philanthropist was determined to make the village a living Williamsburg. He moved historic sites, opened a vista to the harbor, and financed the reconstruction of Stony Brook. The Carriage House is recognized internationally with one of the largest collections of horse-drawn vehicles in the world. Over 50 vintage, transport artifacts, made by both American and European manufacturers, are housed in Stony Brook Museum.

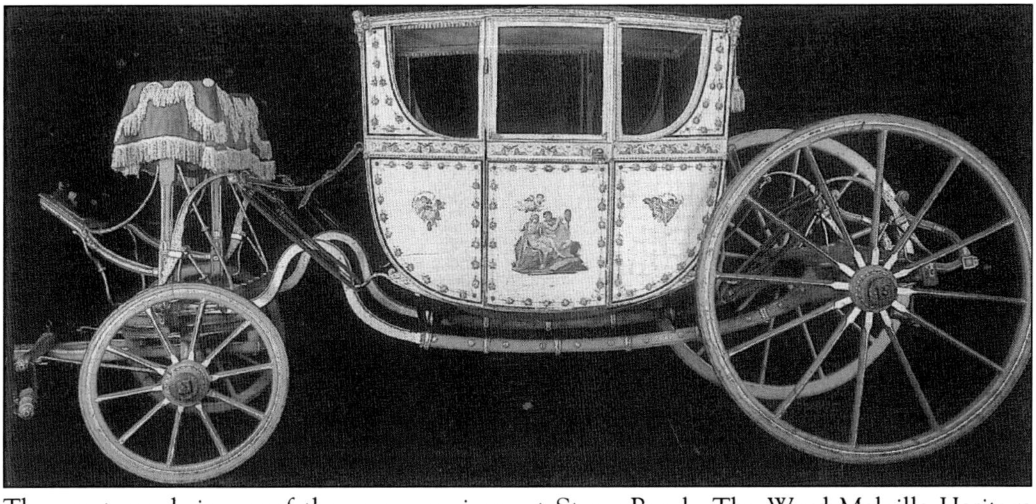

The great coach is one of the many carriages at Stony Brook. The Ward Melville Heritage Organization was formed to honor a man who restored and protected historical artifacts. He created the nation's first planned business center-a charming, colonial, crescent shaped area. It has expanded in response to changing demographics; however, its colonial atmosphere has remained intact. Harbor Crescent, Inner Court, and Market Square include many fine shops of the Fifth Avenue variety.

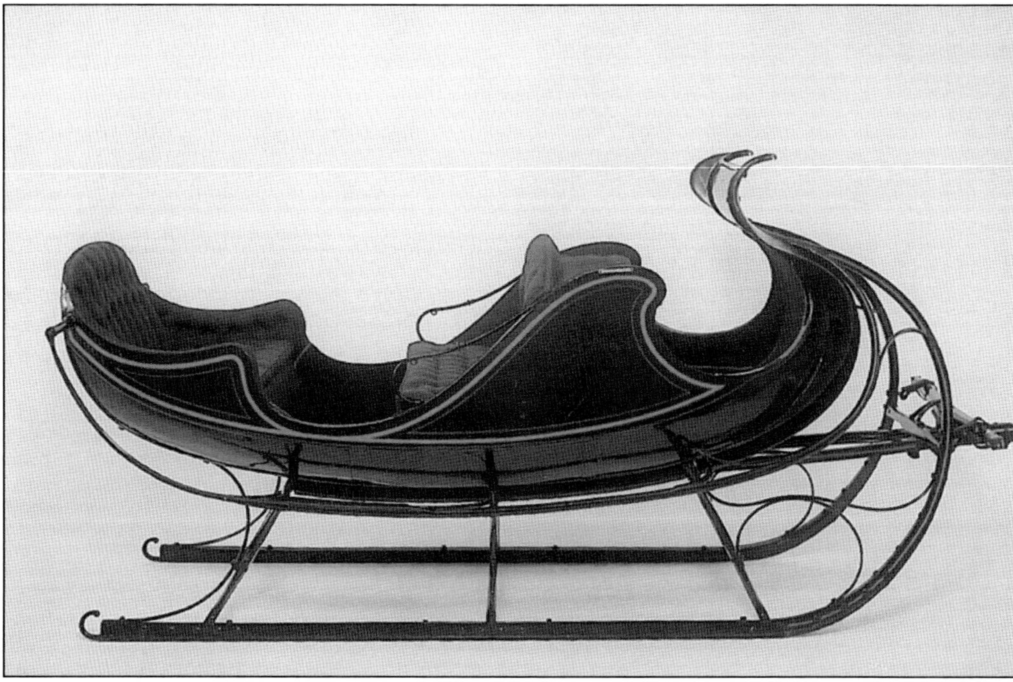

This sleigh of 1880 could well have been used on the roadways of Long Island. A century ago, much winter travel depended on the sleigh. This one is retired to the Stony Brook Carriage House and is on display there.

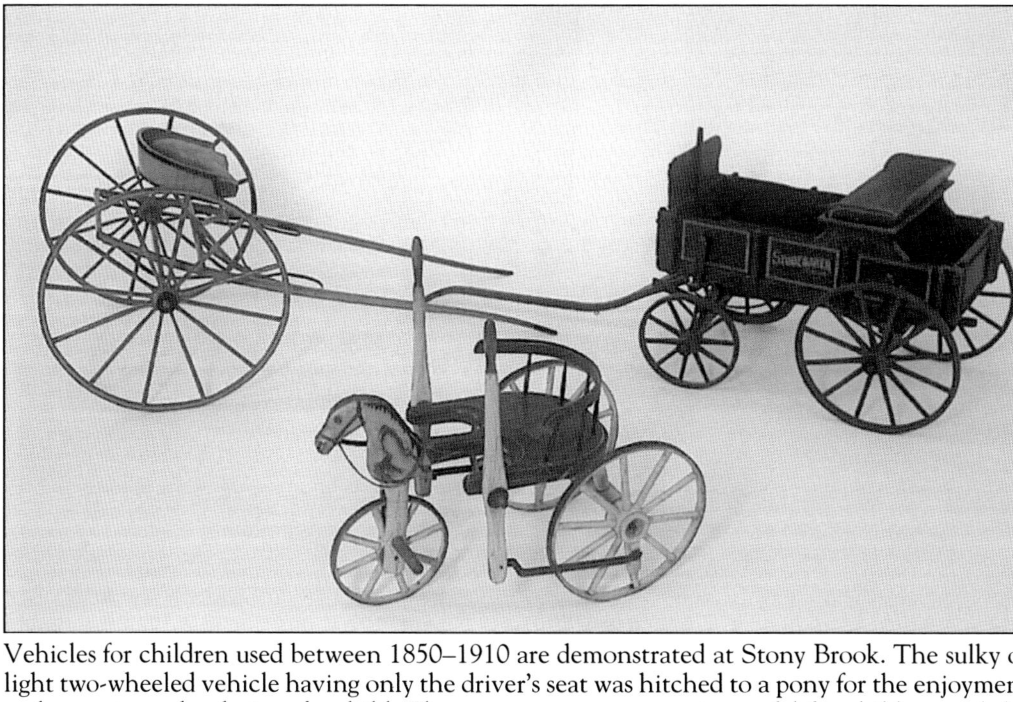

Vehicles for children used between 1850–1910 are demonstrated at Stony Brook. The sulky or light two-wheeled vehicle having only the driver's seat was hitched to a pony for the enjoyment and sometimes the duties of a child. The express wagon was quite useful for children and the tricycle or hobbyhorse was made of wood. The child's father often constructed all three.

Three

Wooden Age Technology

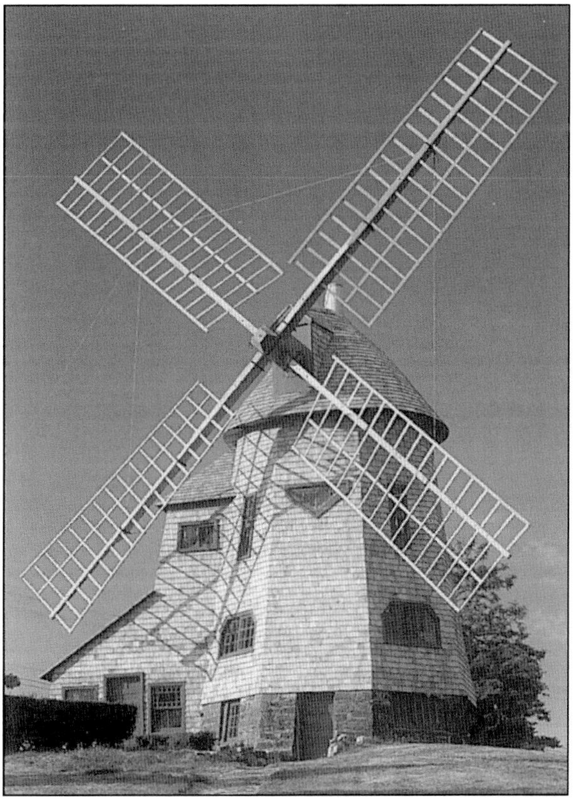

Mill Hill Windmill was built in Southampton where it stood on present day Windmill Lane. It operated on the same site for 178 years. In 1890, it was moved to a site overlooking Peconic Bay on the north with Shinnecock Hills on the south. This land was once an Indian hunting ground. The windmill became a landmark for mariners and villagers.

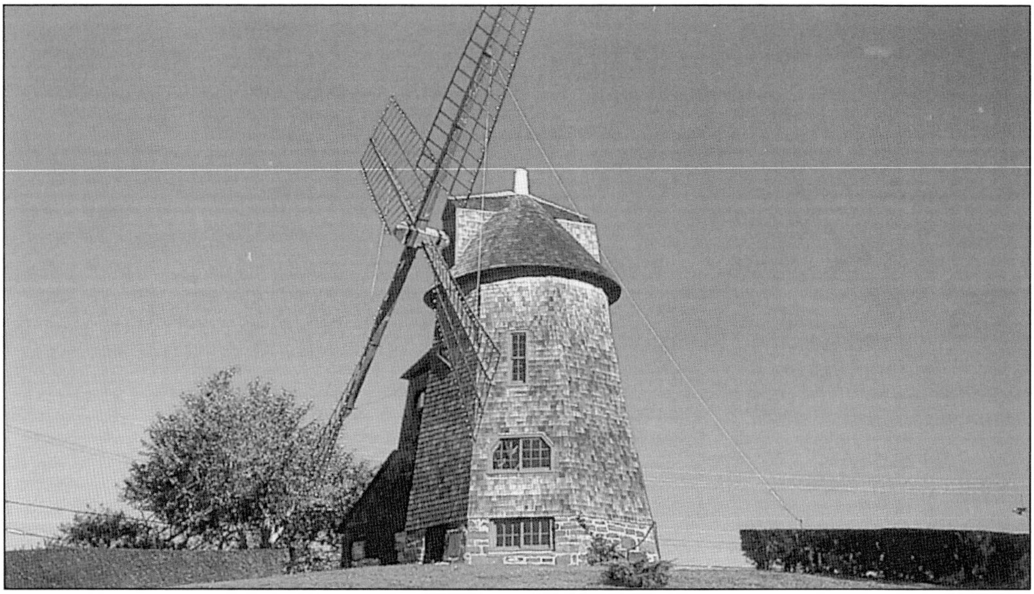

Arthur Clafin, textile manufacturer and financier, purchased the Windmill at Mill Hill. He built a square "cottage" near the windmill and furnished his three-story, 30-room cottage from the furnishings of an European castle. Guests were entertained in the mill. It became a conversation piece. Clafin's little daughter enjoyed using the historic windmill as a playhouse. Clafin called his estate Heathermere. Following World War II, a couple bought the mill and opened Tucker Mill Inn as a restaurant.

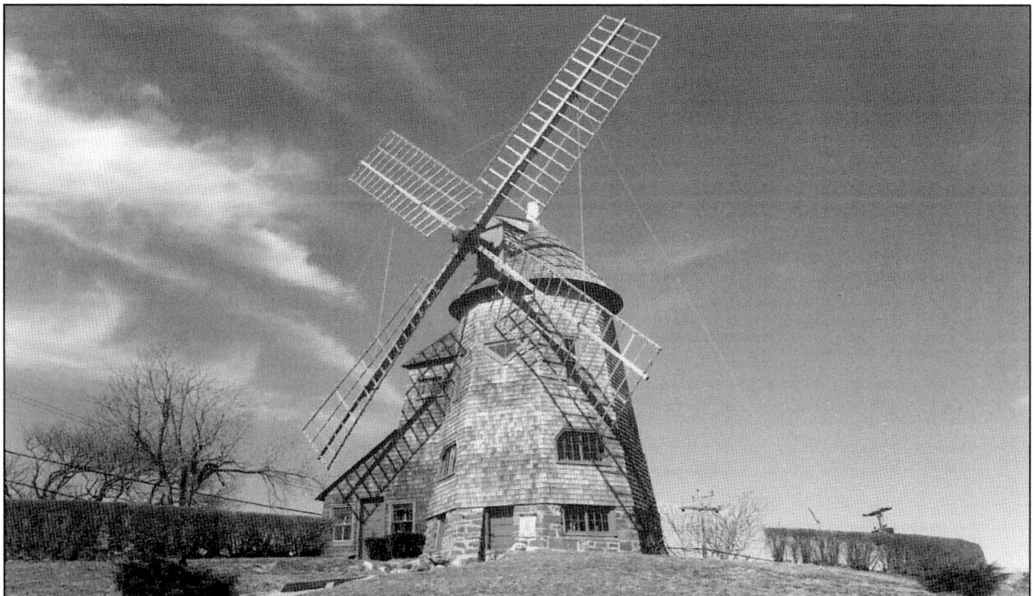

In 1963, historic Mill Hill Windmill embarked on a new career. Clafin Place, later Tucker Mill Inn, was the chosen site for a new institution of higher learning in the heart of the Hamptons. In March 1963, 135 acres were purchased and that fall classes began at Southampton College. The entire school was housed in Heathermere during that first year. Long Island University was later invited to become a partner in the venture.

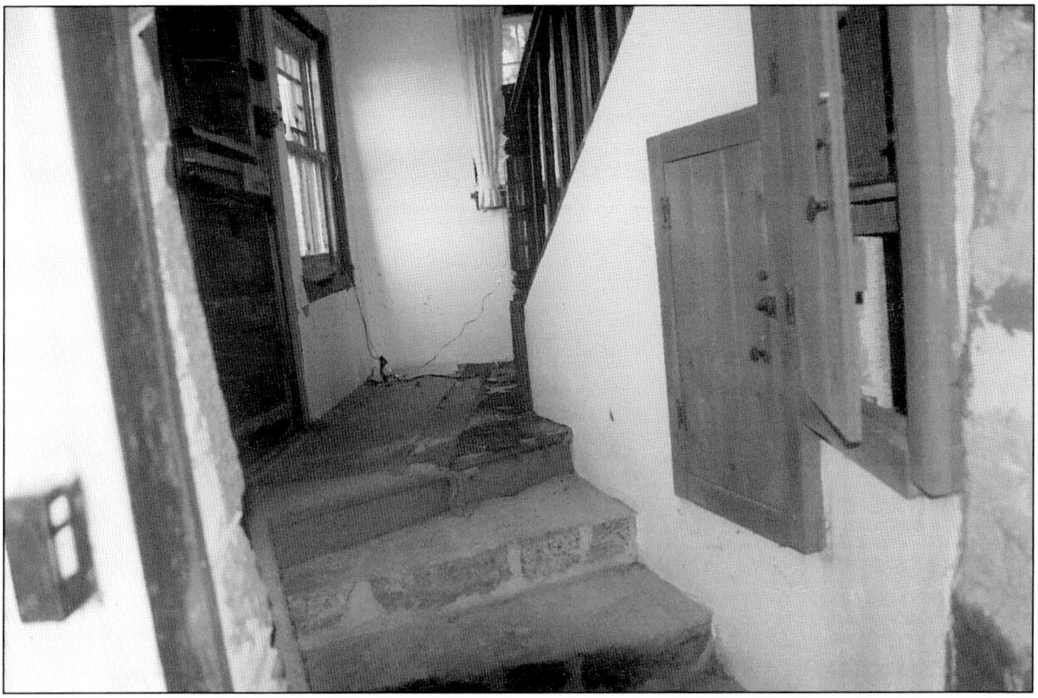

The interior of the College Windmill became a gathering place for faculty and students alike. The winding staircase leads to the upper floors than once housed the machinery for grinding grain. The second level became a cozy kitchen with a circular table surrounding the shaft. The shaft once gathered power from the huge sails outside. An anchor served as a handrail. The third level housed a study and office.

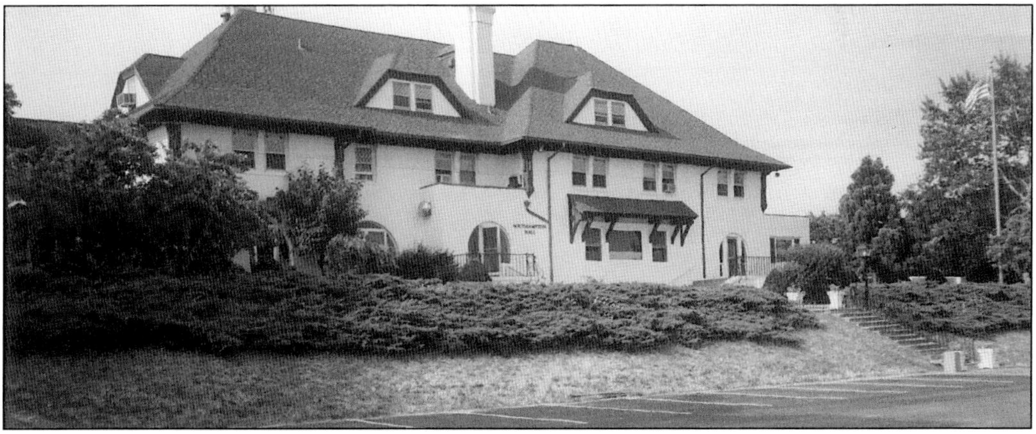

The mill has always been revered at Southampton College and Long Island University. The school has academic divisions leading toward degrees in business, fine arts, the humanities, natural and social science, and teacher education. The Marine Science program is well known for its own fleet of vessels, which includes a 44-foot ocean liner. During the golf tournaments in Southampton, the dormitories are rented to visitors.

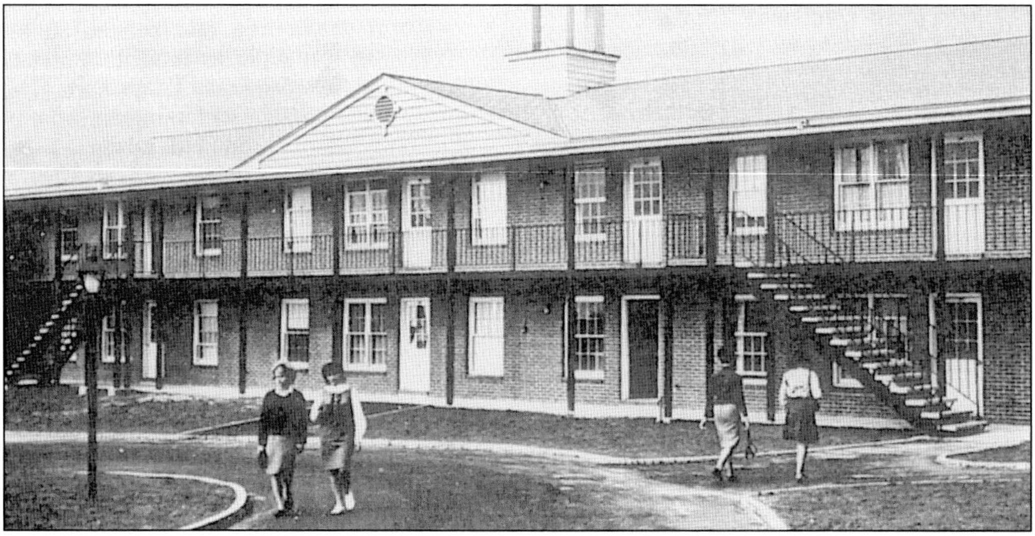

A college dorm complex graces Southampton College. It was a small liberal arts college offering Bachelor Degrees. The Department of Marine Science is included in this ocean-oriented resort area. Southampton has become a multi-campus town with the arrival of Long Island University. The college and university retain pride in the mill. When the athletic team achieves success, a flag is hoisted on the windmill to announce the victory. The college newspaper is *The Windmill.*

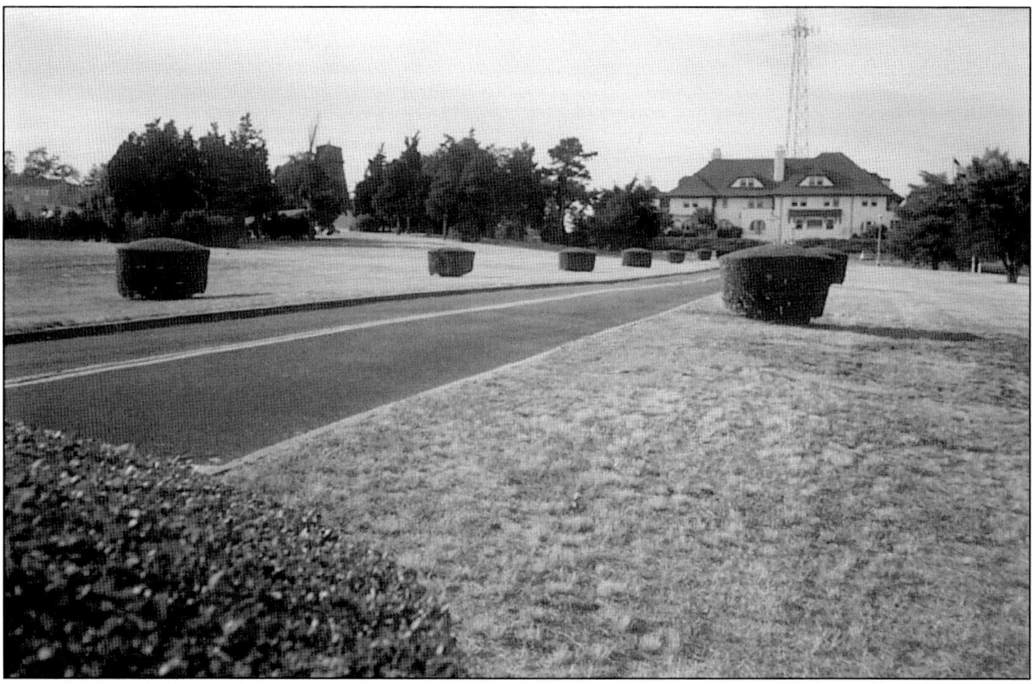

The Clafin Cottage lives on. The winds that once set the 35-foot sails of the College Windmill revolving still blow in from the ocean. The massive wind shaft was hewn from white oak. The turning cap of the mill, the giant chocks, the shaft, and millstones were all part of the mill's restoration. The college windmill frames are now attractive flowerbeds. The venerable 286-year-old windmill is the pride of Southampton College.

The Golf Course Windmill was built in 1916 as a tribute to Charles Blair Mac Donald. He was a New Yorker and the first American amateur golf champion. He held the world's record in 1896 and 1897. In 1906, Mac Donald bought 200 acres at Shinnecock Hills and had the National Golf Course lain out by Seth J Raynor, a civil engineer. The golf links opened September 16, 1911. The mill originally pumped water for a fresh water well at the clubhouse.

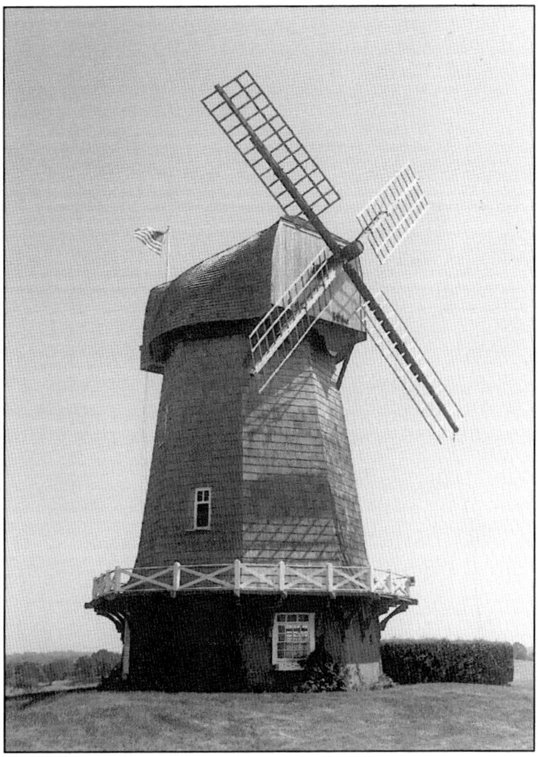

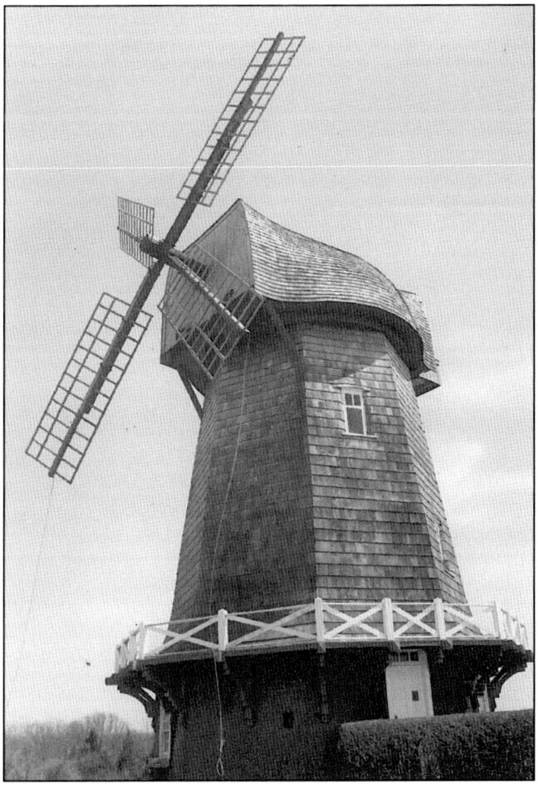

Southampton's Shinnecock Hills became the scene of the first golf course in the United States. The treeless dunes and moors of Shinnecock suggested the landscape of Scotland where the game had been invented. A new and vibrant life was dawning in the Hamptons as it evolved into a resort for the wealthy and affluent of New York. National Golf Links were established at Southampton in 1906. The clubhouse was built in 1912 with plans drawn up and donated by Jarvis Hunt.

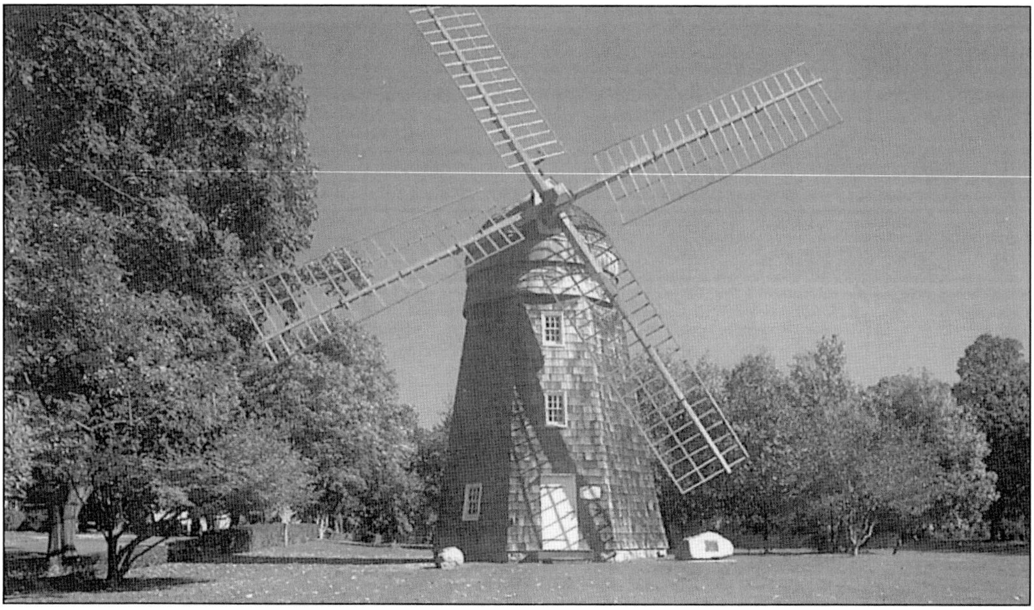

The Beebe Windmill was built in Sag Harbor in 1820 for Lester Beebe who was a retired whaling captain and shipbuilder. Constructed by a local woodworker and millwright from Amagansett, it was the tallest structure in Sag Harbor. It was a custom to raise a flag on the mill whenever a whaling vessel, returning from afar, entered the bay. The Beebe Mill demonstrates American technology in transition; it was later converted to steam power.

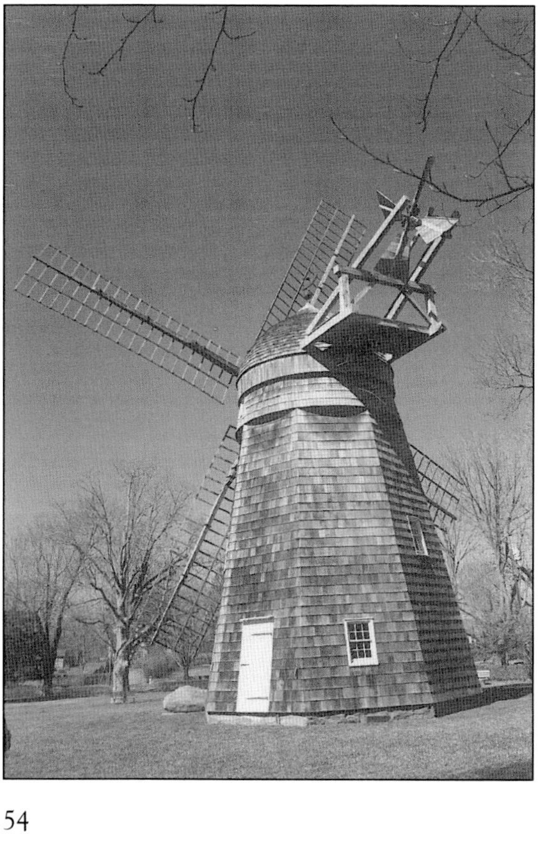

The Beebe Smock Windmill is one of only a few remaining from the days when wind and water were the only source for grinding grain into flour. Today it stands on the Village Green in Bridgehampton, the last of many sites it has occupied in its long, adventurous history. The fantail automatically revolves the cap, which rotate the sails. The smock windmill has an unusual ogee. It keeps the mill winded or facing the wind. It has double-sided common sails.

The Beebe Windmill began its travels following Beebe's death. It was bought by Rose Gelston who had it moved to Bridgehampton and placed on high ground north of the Presbyterian manse. It operated here for a half century under a sequence of various owners. In 1882, James Sanford purchased the mill and moved it to a site near the railroad. He later installed a steam engine in the mill.

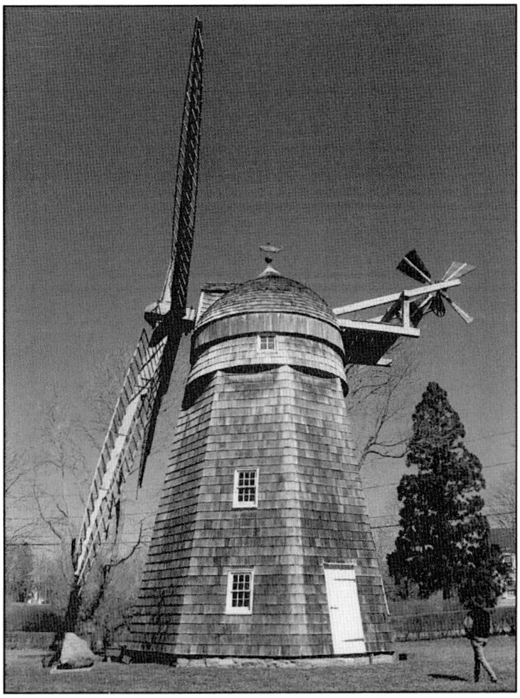

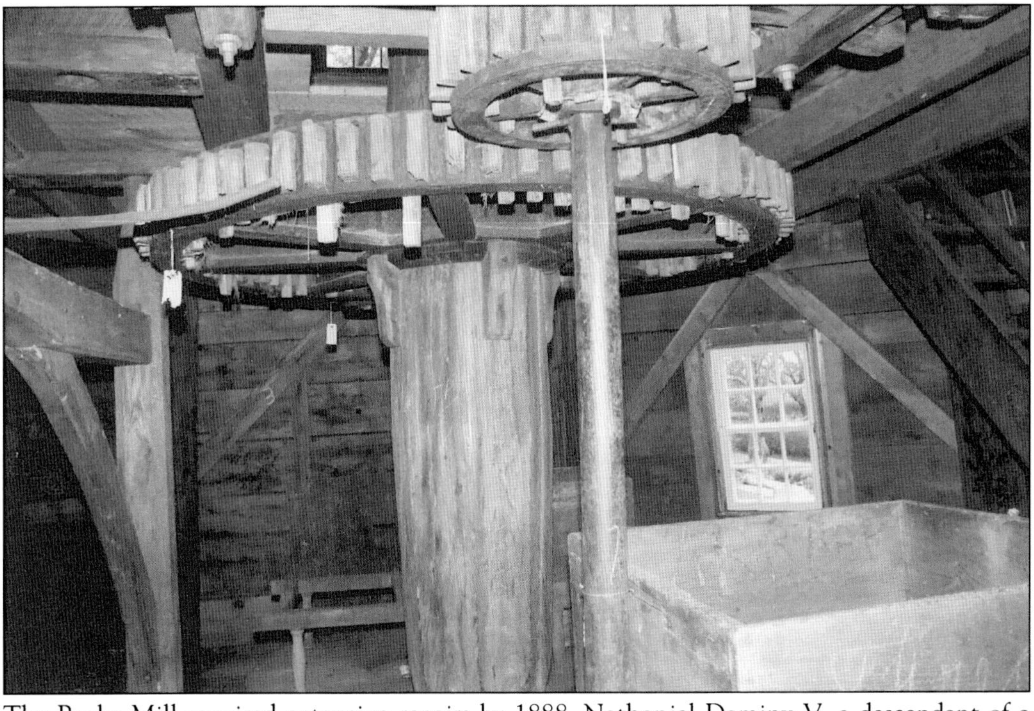

The Beebe Mill required extensive repairs by 1888. Nathaniel Dominy V, a descendant of a long line of noted millwrights on Long Island, did the repairs to the mill. The following year, it was moved to the north side of the tracks. There, under the direction of the Bridgehampton Milling Company, it operated successfully for the next two decades. In 1915, John E. Berwind bought the historic mill and moved it to The Grassy Green beside his summer estate, Minden.

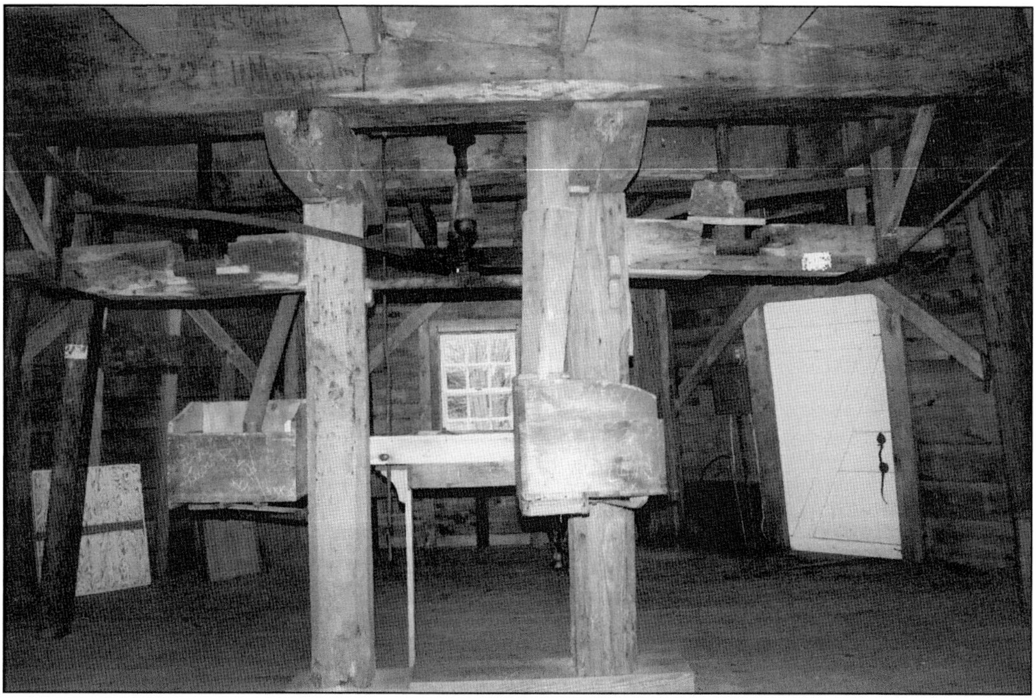

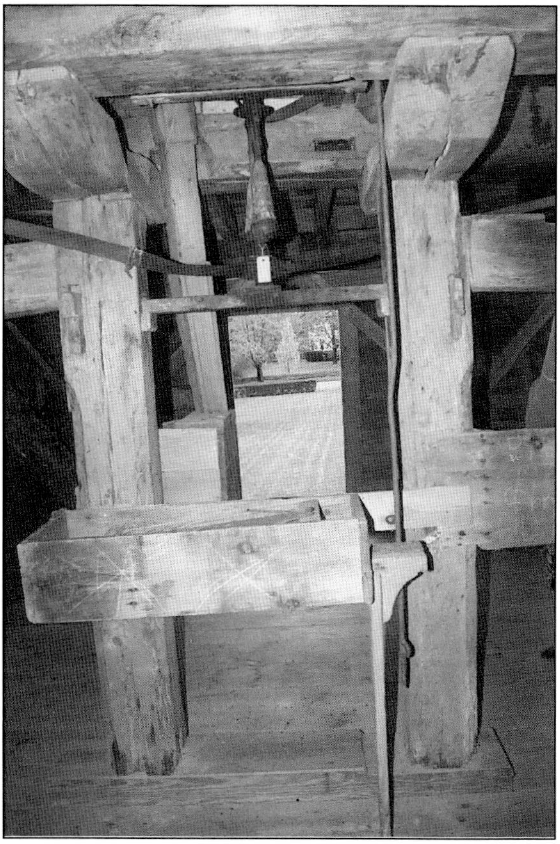

The Beebe Mill has a stone foundation and originally had a four-story stationary octagonal tower housing the millstones. It initiated cast iron gearing and marked the transition from wooden technology into the era of iron machinery. Note the grinders and meal bin for receiving flour. Here the flour can also be bagged. The wooden steps on the right are where the miller took the grain to begin its journey through the grinders.

The interior of the Beebe windmill provided for cleaning grain as well as grinding it. As the grain began its journey, it was cleaned in the scouring machine that blew air through it. It was then poured through a hole in the running or capstone to the grinding surfaces and stationary bed stone. The flour is carried to the edge of the stones by furrows and into a chute below. It is then sifted and bagged.

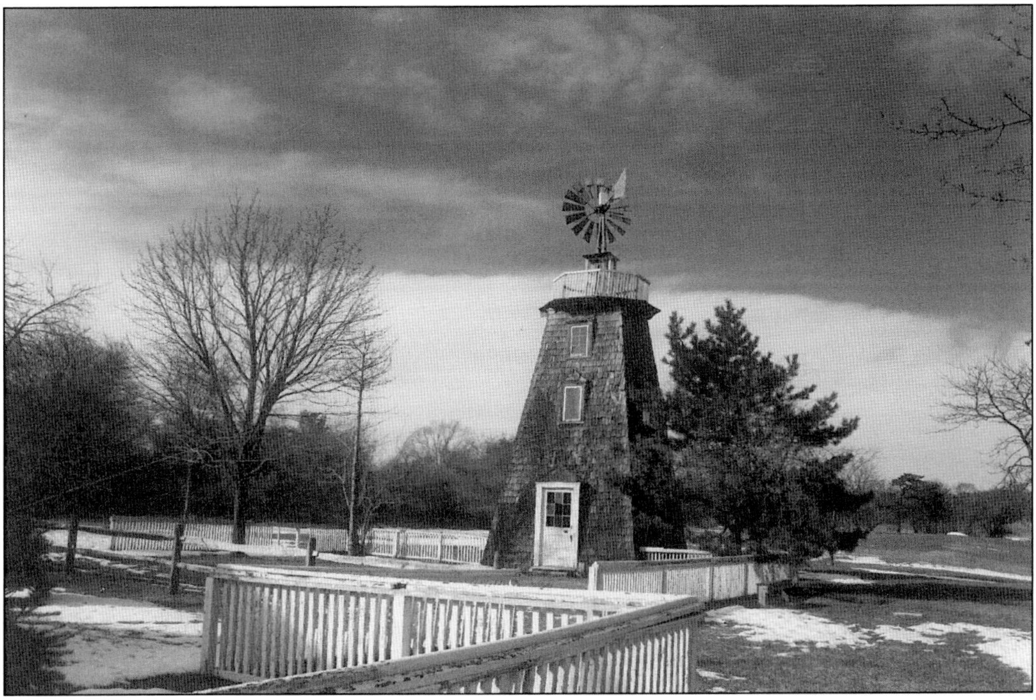

The American Farm Windmill was built in 1890 on the Powell farm in Sayville. South Shore estates used wind power to pump water, irrigate their land, water the livestock, and pipe water into their homes. In 1914, it was bought by Arthur Udell who moved it with his wagon and oxen to Bohemia. He constructed a cement foundation and centered the mill structure on it. In 1975, it was moved to the Town of Islip Grange as memento of Americana.

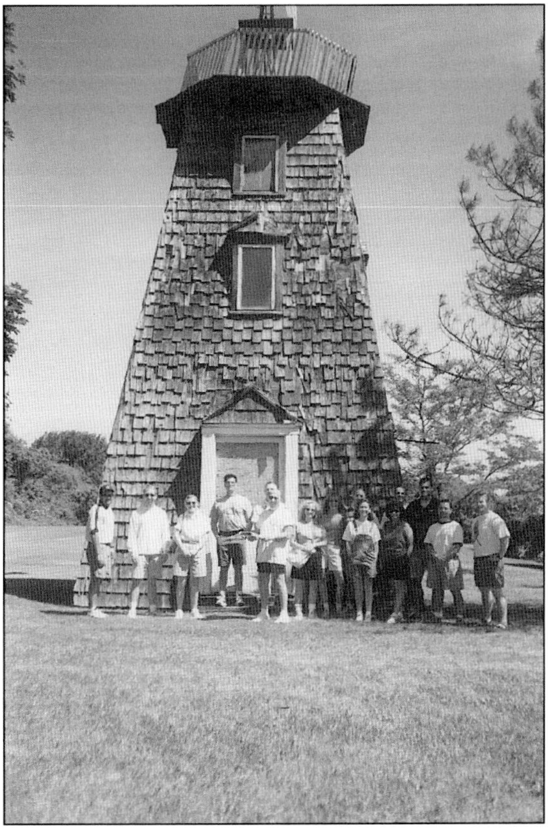

Gerald Leeds, a miller and teacher takes a group of teachers on a tour to the Islip Grange Mill. This gives teachers an insight into the wonders of wooden technology. The Town of Islip Grange is a 12-acre site set aside in 1974 as a repository for 18th and 19th-century structures that were threatened with demolition. Some of the structures reflect rural village and pre-Civil War days. The American Farm Mill is among the buildings preserved here.

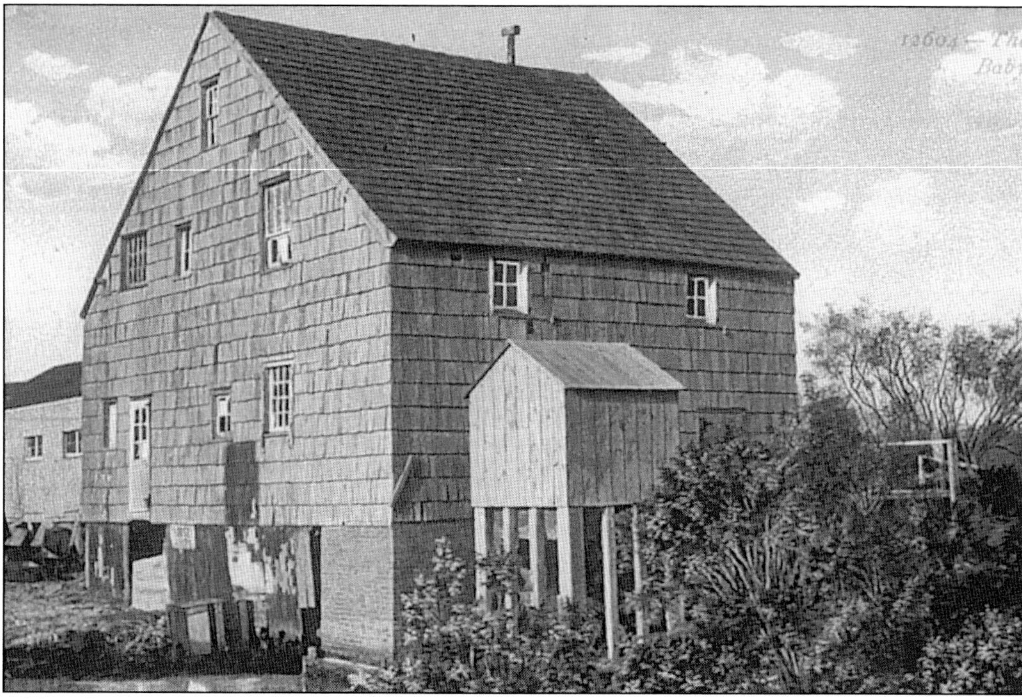

Old Mill in Babylon is also known as Hawley's Mill. It was operated by Nathaniel Oakley. The source of water was Sumpawams Creek that flowed from the Great South Bay into Oakley Pond. Nathaniel Oakley was famous for his originality in dress. He wore a smock frock that fell below his knees with a high black, beaver hat. These were well-covered in white flour dust. In later years, his son, Eliphalet, changed the mill from a gristmill to a sawmill.

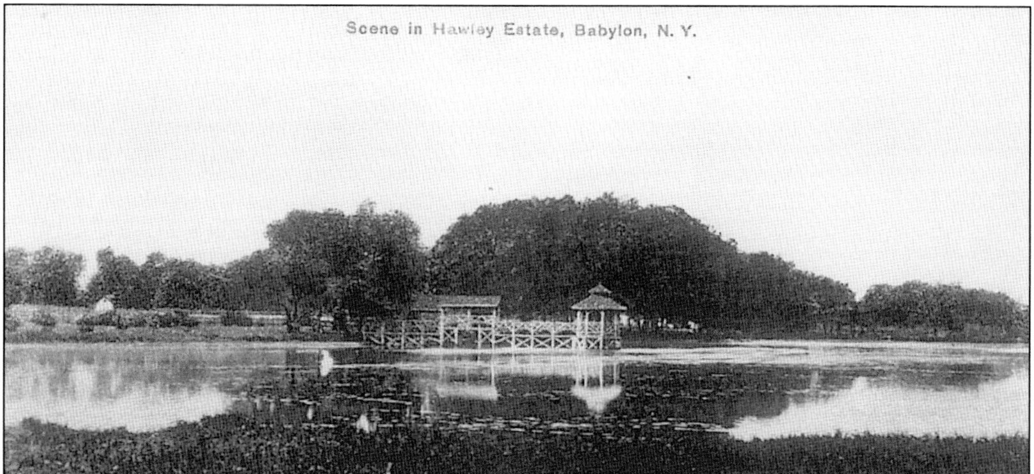

The Old Mill in Babylon made history with the Babylon Electric Light Company. The mill on Sumpawams Creek generated the first electricity on the South Shore. On November 24, 1886, electricity illuminated eight stores and three streetlights. A few years later, a larger site was found and the mill became a buggy whip factory. Hawley, owner of the Sutton Estate replaced the mill, with a scenic waterfall on the eastern approach to Babylon.

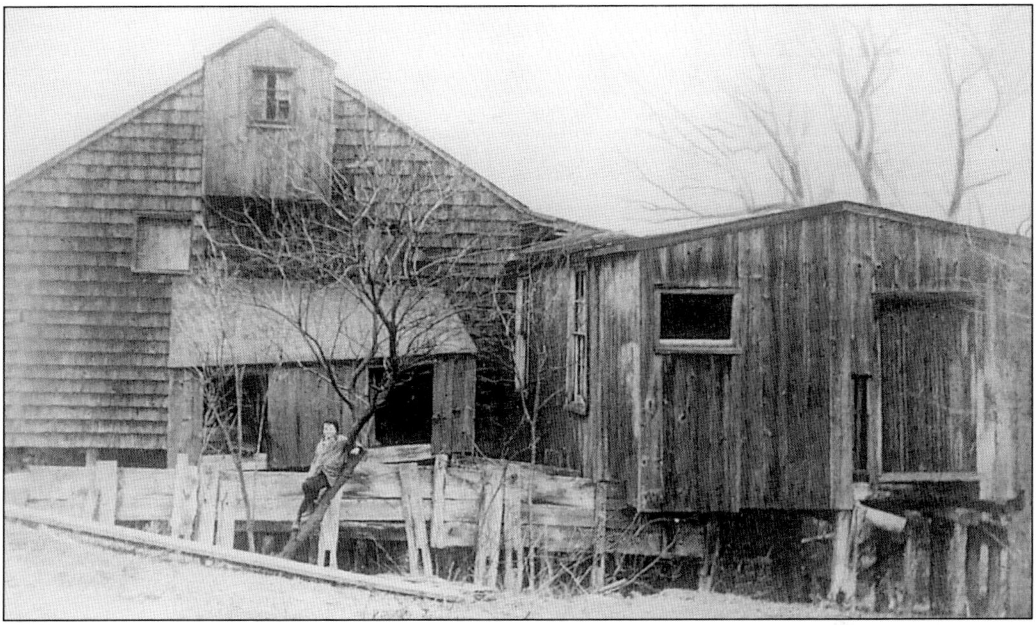

Southard's Grist Mill in Babylon was operated by Joseph and James Southard. It ground corn and flour for Babylon and the nearby communities. R. Villefeu was a devoted miller. According to the *South Side Signal* of Saturday, June 5, 1886 ". . . he has returned to his old love, the Southard flouring mill, and is now running it, having leased it for a term of years." New machinery had been installed. Southard's Pond has become part of the Long Island State Park Trail.

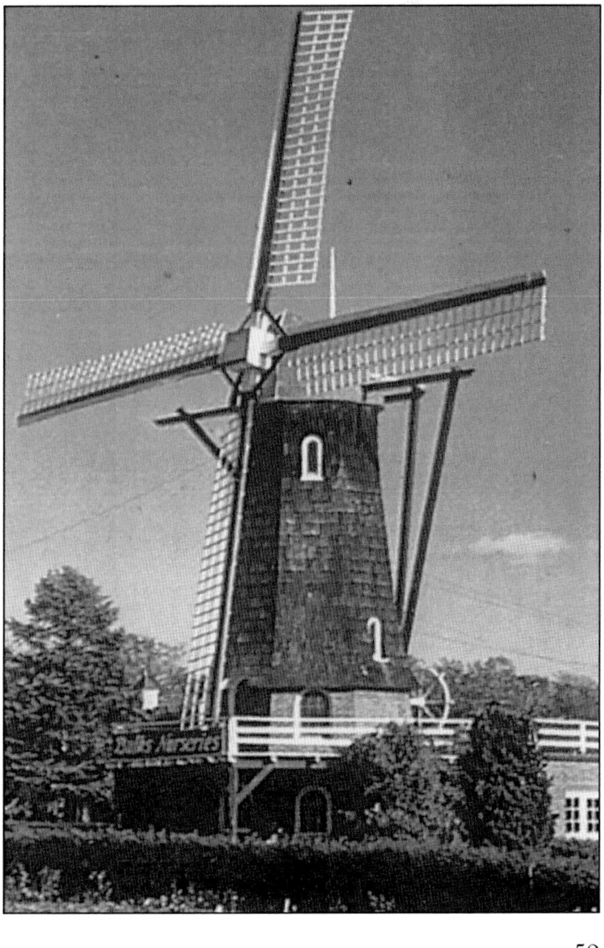

Bulk's Windmill in Babylon was a decorative attraction to his garden center. In 1929, he established a business on Montauk Highway. He constructed a four-story windmill that was an actual reproduction of a Dutch mill in his hometown on the Rhine in Holland. It created an atmosphere of enchantment in the area. Bulk's offices were housed in the mill. Sturdy and dependable like the mill, were synonyms for Bulk's service.

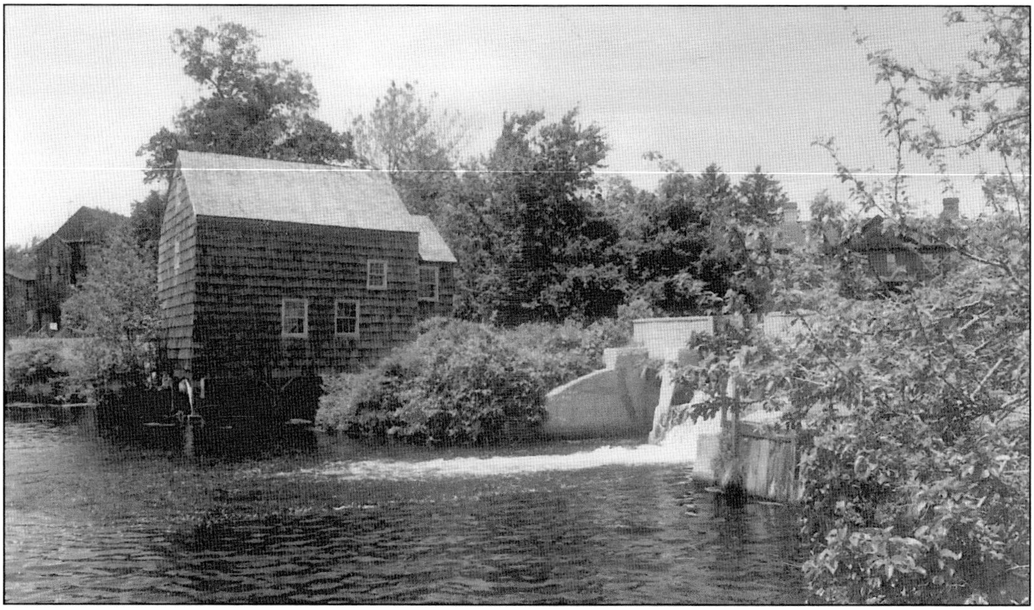

The Nicholl family built the Oakdale Grist Mill in 1740. In 1683, William Nicholl, founder of the town of Islip, bought an 8-by-10-mile tract of land from the Connetquot Indians. Nicholl called his manor Islip after his hometown in England. The family settled on the south shore and the mill became a vibrant part of the economic and social life. It survived as a horizontal drive mill until 1878.

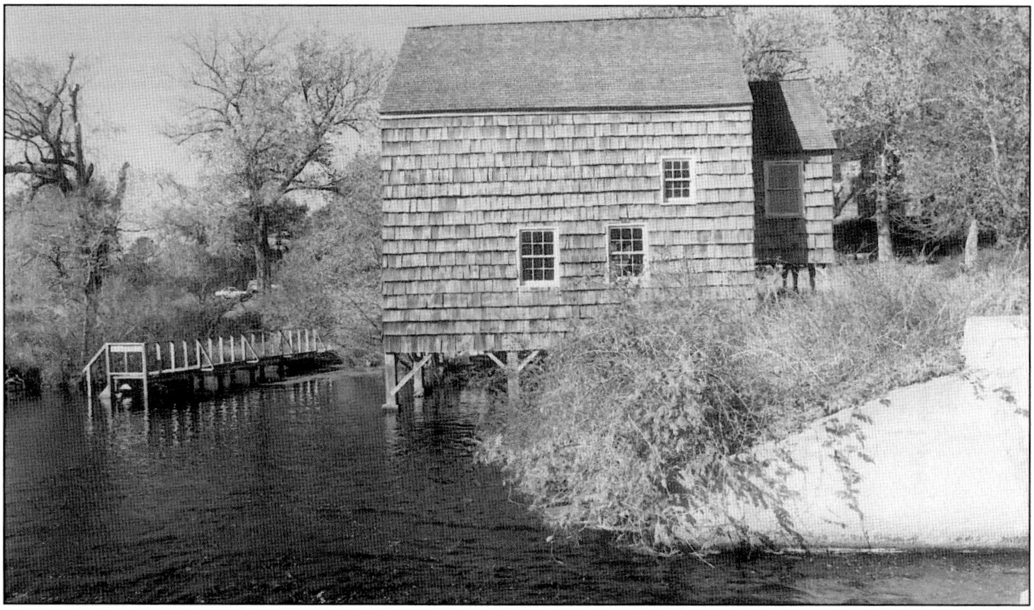

Nicoll's Mill, on the southwest banks of the Connequot River in Oakdale, served the needs of local residents. In 1866, The Southside Sportsmen's Club took possession of the mill. It was left idle for a century. Weather shingles protected the renowned mill and most of it survived. The Long Island State Park Commission took over the mill and the Vanderbilt Historical Society of Dowling College restored the historic mill.

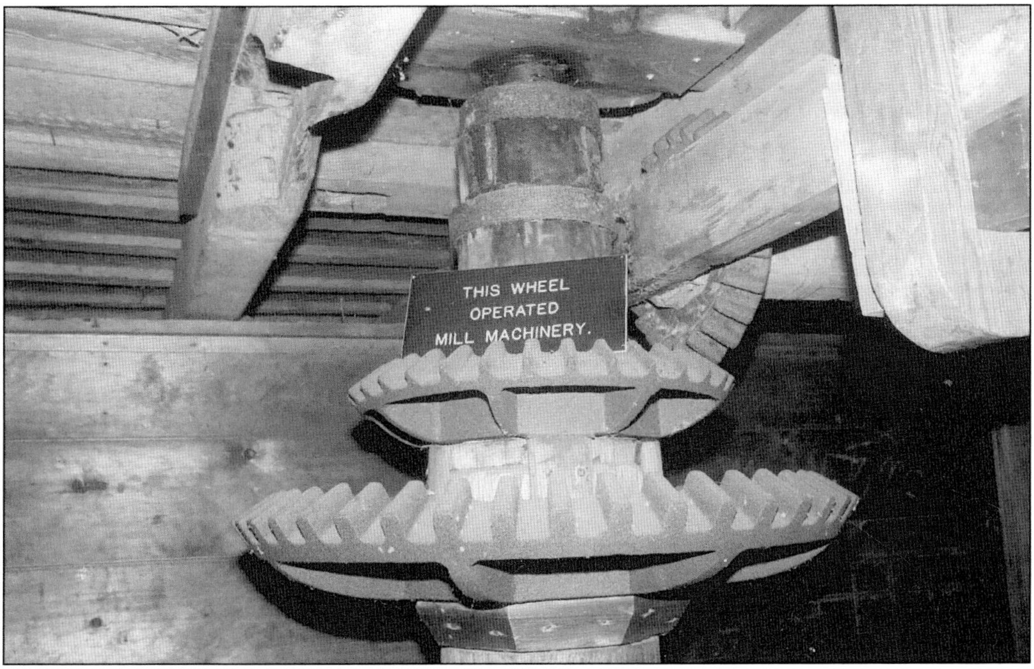

The horizontal water wheel at Oakdale Mill is different from the vertical wheel of other mills. The force of rushing water pushes the paddle away from the source and turns the millstone directly. Horizontal wheels were popular until technology advances progressed and greater power was generated with a vertical drive wheel. The Oakdale mill featured a tendering mechanism to adjust size and flow of water whereby flour flowed from spout to bin.

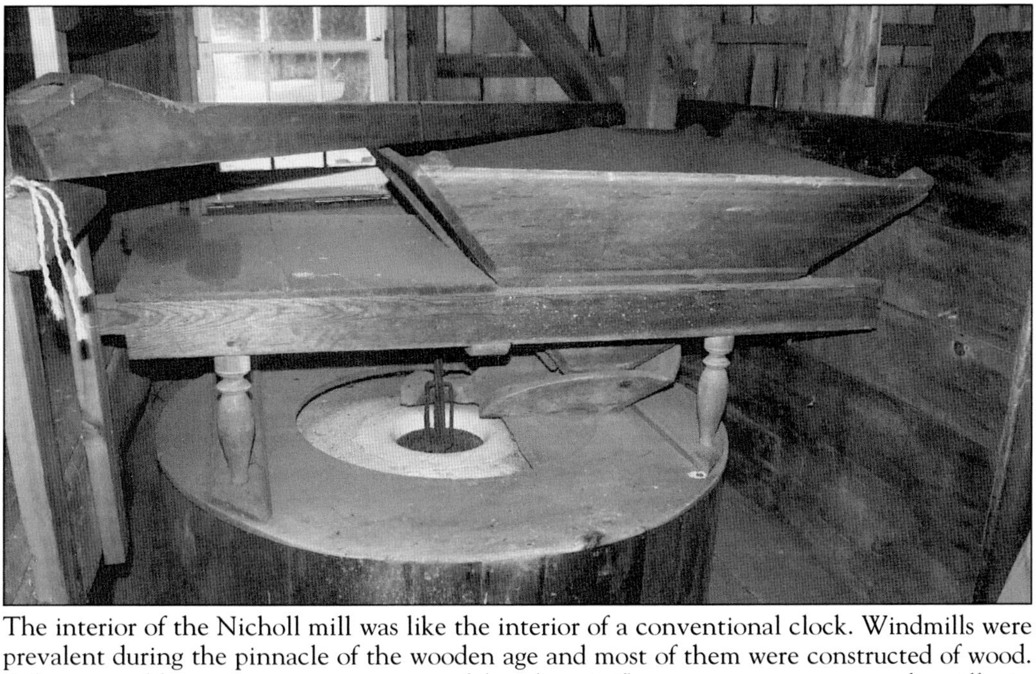

The interior of the Nicholl mill was like the interior of a conventional clock. Windmills were prevalent during the pinnacle of the wooden age and most of them were constructed of wood. A farmer could expect to get ten percent of the wheat in flour, ten percent went to the miller in payment, and a certain percentage was lost to the machinery.

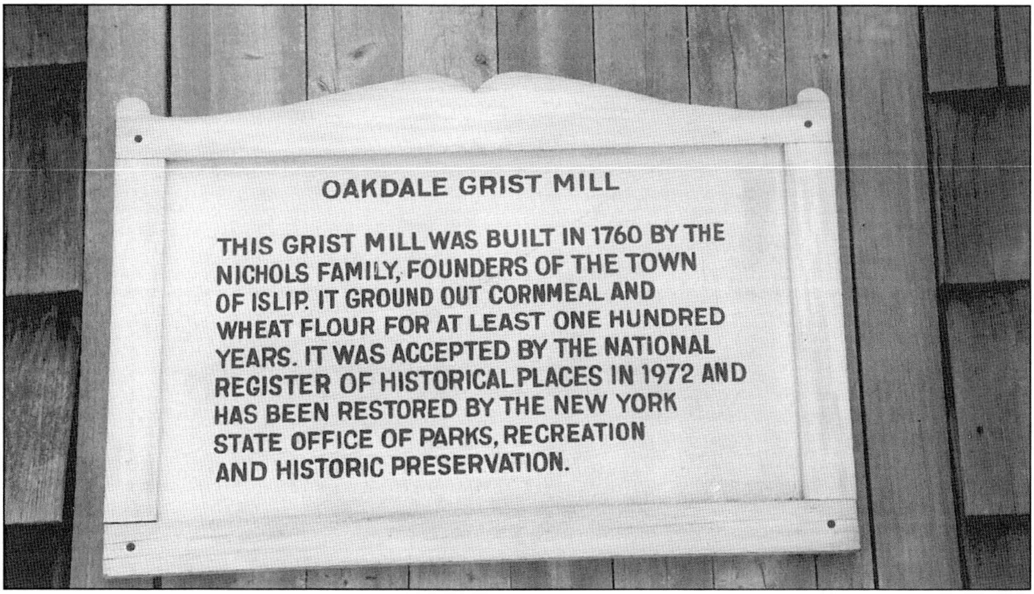

A plaque on the Nicholl Grist mill attests to its authenticity. The mill has been modernized and modified many times but the drive mechanism was never changed. The Long Island State Park and Recreation Commission has refurbished the exterior. In 1972, it was placed on the National Register of Historic places. It presently stands in Connetquot State Park, Oakdale.

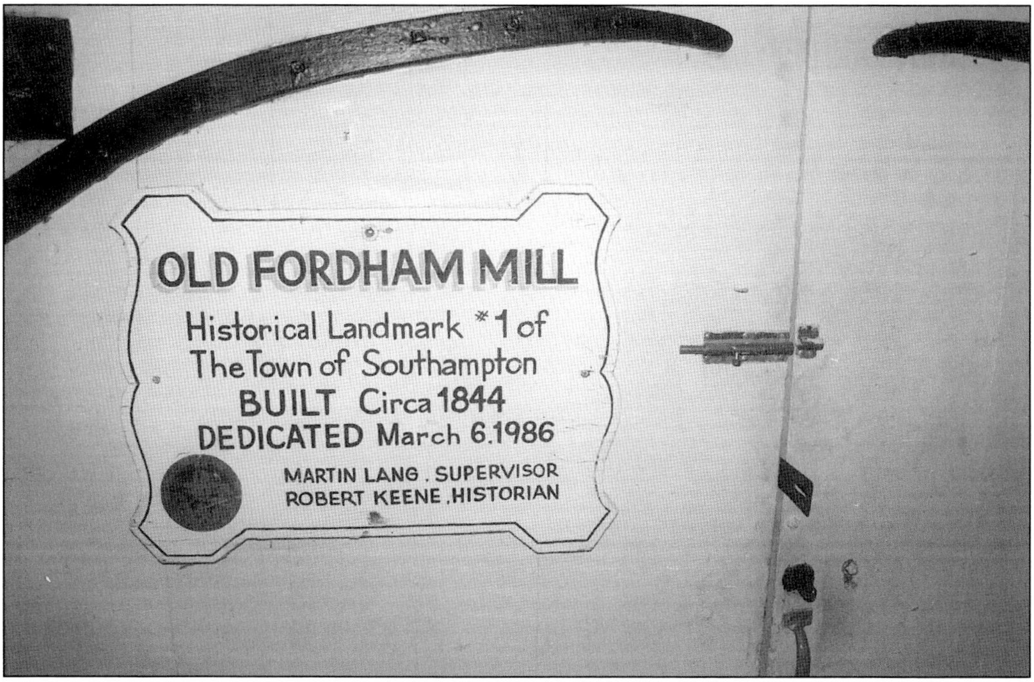

Tuttle Fordham Mill was built in 1859 in Speonk. Daniel Tuttle constructed the mill to house the carriage manufacturing business he had established in 1844. The mill was equipped to saw, bore, drill, and do wood work as well as metal. This was powered by a huge water wheel beneath the building. It was converted to electricity in 1911 where a sawmill and lumberyard operated for many years. The mill later served as a furniture store.

Bourne's Windmill was situated on the Bourne farm along the shores of the Great South Bay. It was a tall, stately building that serviced the entire estate. Frederick Bourne, president of Singer Sewing Machine Co. acquired a 3,000-acre estate in Oakdale. His son, Arthur, established the farm in 1911. Bourne's mansion house is now LaSalle Military Academy, staffed by the Irish Christian Brothers.

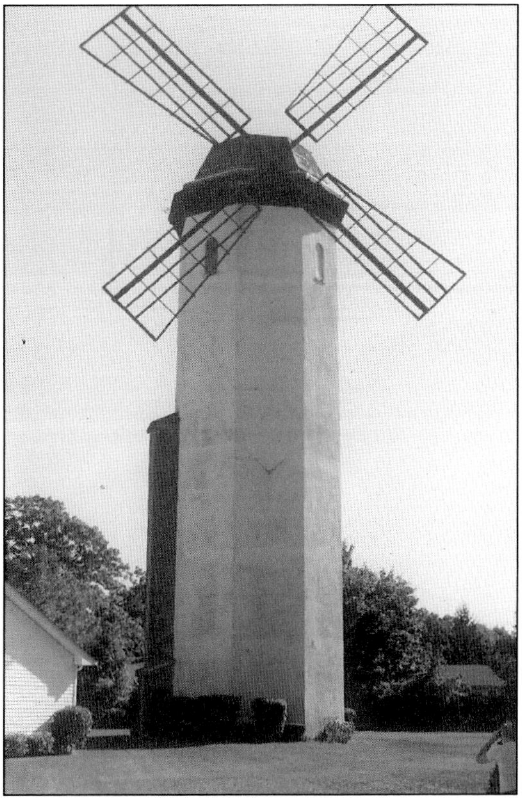

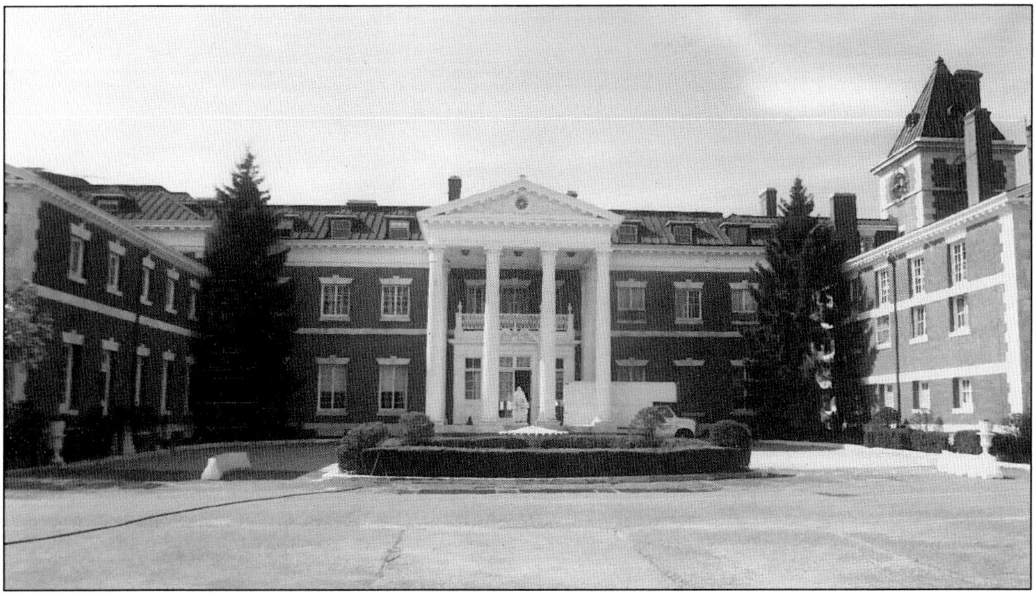

La Salle Military Academy in Oakdale was once Indian Neck Estate. Frederick Bourne, president of Singer Sewing Machine, built a spacious mansion fronting the Great South Bay in Oakdale. Called Indian Neck, it was completed in 1897. Bourne was an avid boatman. The family lived here for many years. In 1926, the mansion was sold to La Salle Military Academy. The school continues to operate and has added elementary classes in recent years!

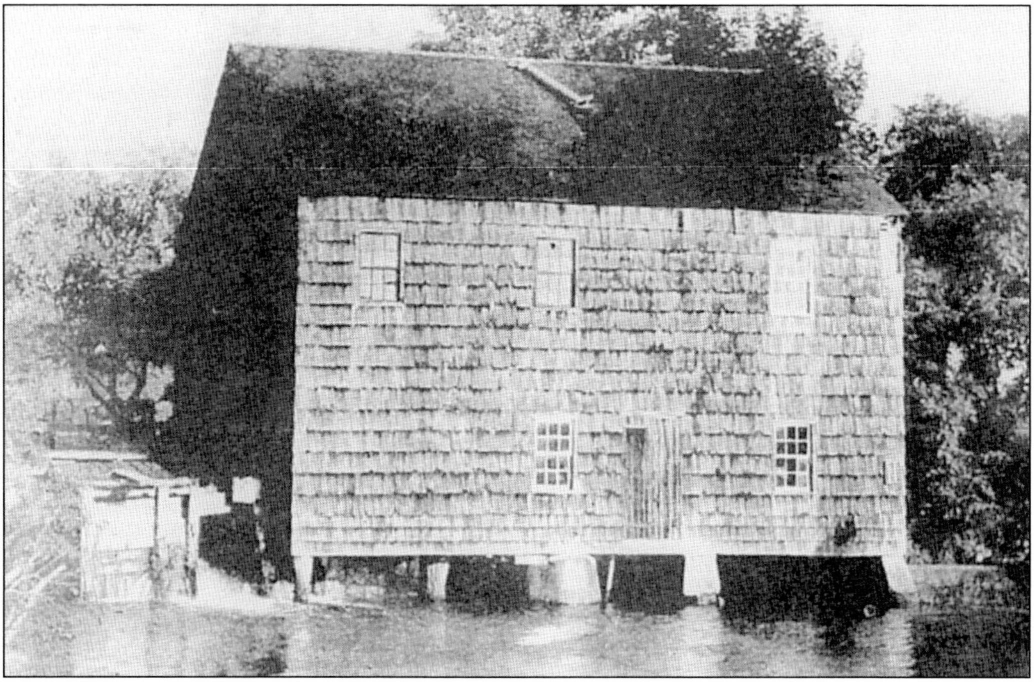

South Haven Grist and Sawmill on Carman's River dates back to 1745. It was here that Daniel Webster caught a 14-pound trout in 1835. A pattern of the celebrated fish was carved on cherry wood as a weather vane for the nearby Presbyterian church. Samuel Carman's tavern was a prominent place to hold political meetings and elections. Prestigious men lodged here to discuss problems of the day and townspeople gathered to hear the latest news. Webster dubbed Long Island Sound "The American Mediterranean."

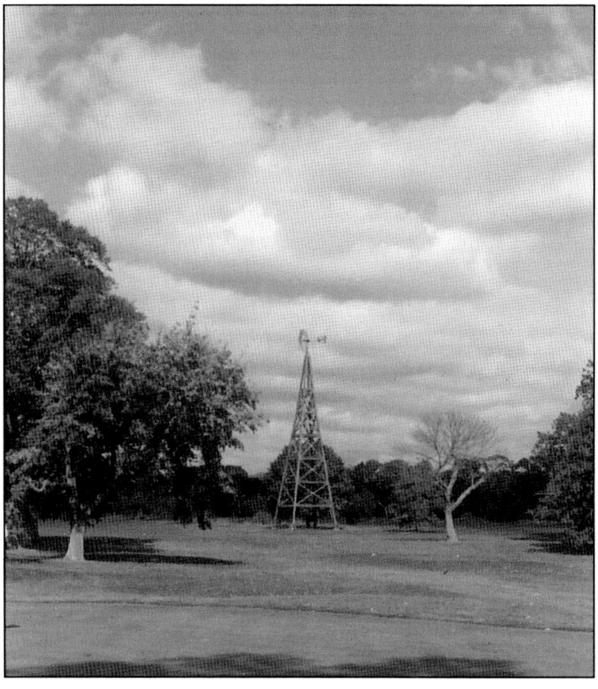

This windmill was part of the Southward Ho Country Club in West Islip. The main building dates back to 1875 when it was built under the name "The Oaks." In 1925, the vacant property and land were deemed feasible as an 18-hole golf course. In the 1950s, the house was remodeled and the upper floors removed. The grand hall remains and the Southward Ho has become an exclusive restaurant. The golf course remains active today.

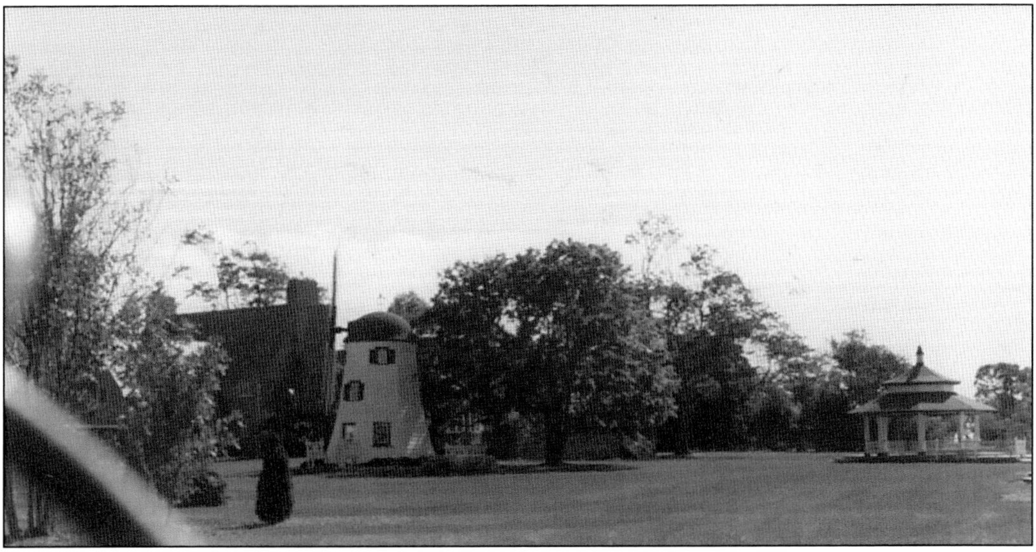

The Conover Windmill was built in 1880 in Bayshore. Mills were part of many Long Island estates during the Gold Coast era. Daniel Conover was a developer who built cottages on Saxton Avenue then rented them. His Victorian house was duplicated many times with its mansard roof and wrap-around screened porch. He improved roads and dredged creeks. His little mill served to grind the grain necessary for the family needs.

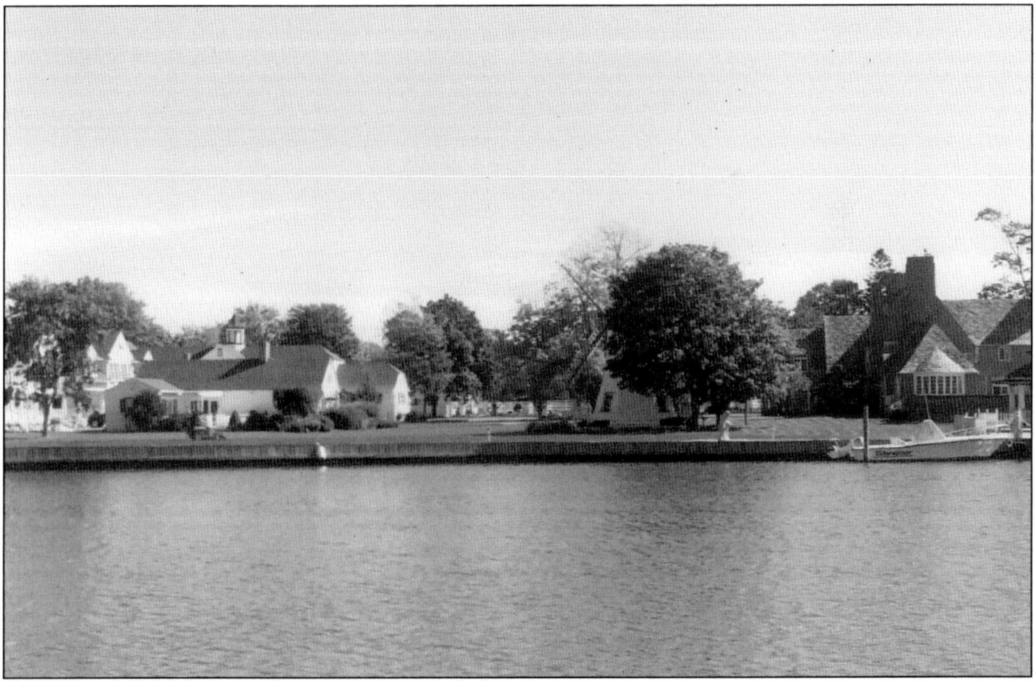

In this picturesque setting, Conover's windmill is partially hidden by the tree. It stands on the banks of a peninsula or inlet known as Awxia Creek that juts in from the Great South Bay. During the era of Gold Coast estates, from the 1880s through the first World War, many tycoons settled along both shores of Long Island.

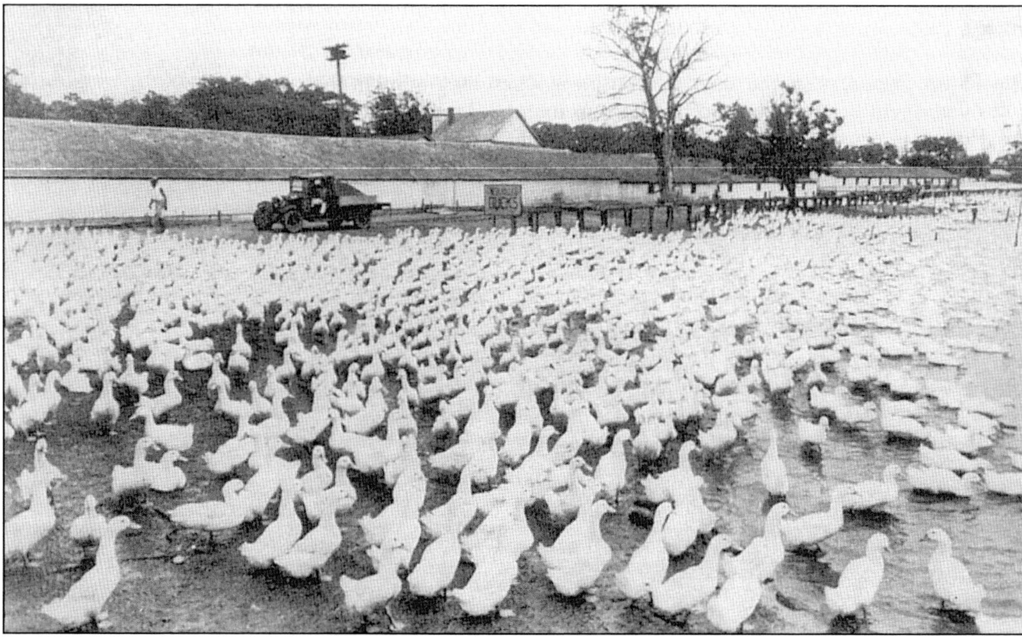

Duck farms are numerous on Long Island. The humid climate, abundance of water and sandy soil are conducive to duck breeding. Three fourths of the duck population in the United States trace their ancestry to Peking, China and live on Long Island. In 1873, a New York merchant James Palmer left Peking with 25 ducks. Four months later nine ducks survived the trip and arrived in New York. These are the ancestors of all Long Island ducks.

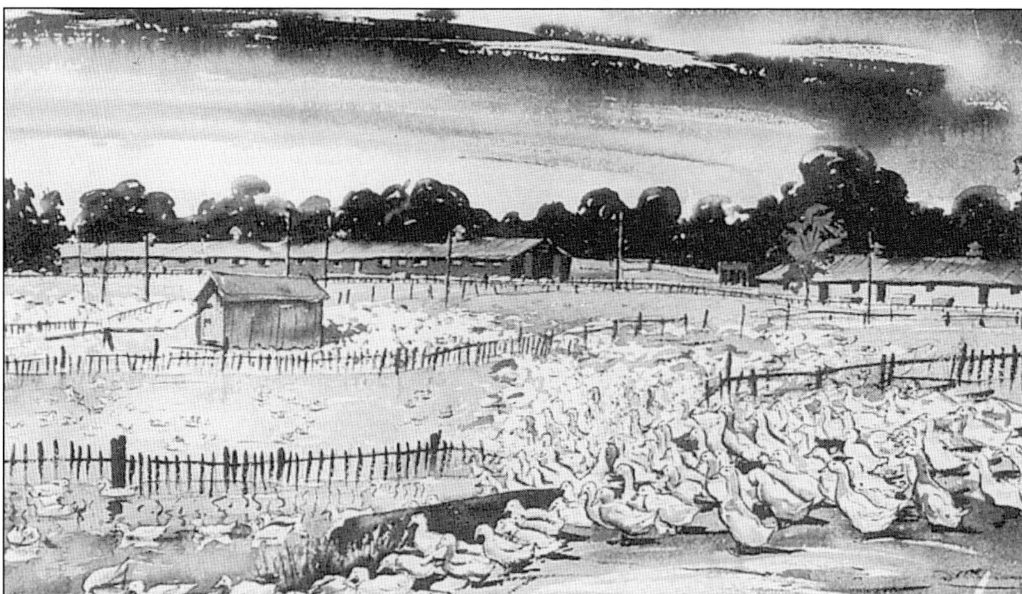

Ducks became a marketable product almost immediately. Duck farms sprang up all over eastern Long Island. A.J. Hallack's farm alone produced 250,000 ducks per year. Ducks mature quickly, lay about 150 eggs per season. Incubators were invented in 1870. A rainy day is not good for ducks. Young ducklings are endangered by rain due to lack of feathers. Today there are processing plants in Eastport and Riverhead.

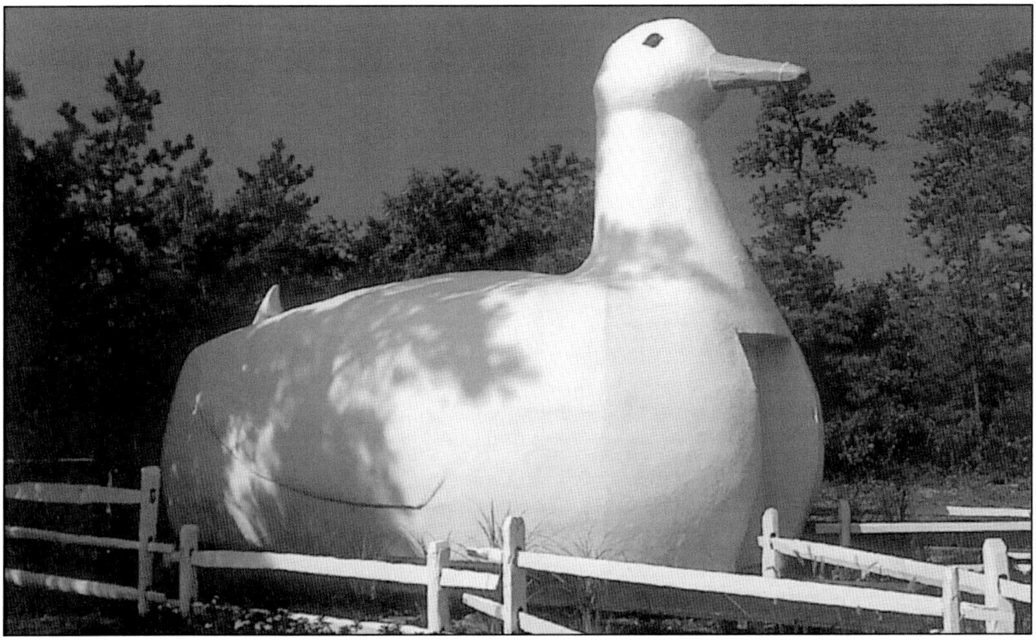

The Big Duck was originally a retail outlet in Riverhead. This duck was built in 1930 when competition was keen. Gimmicks to lure customers were prevalent. Martin Maurer consulted the Collins brothers, who were designers of props and scenes for New York theaters. They built the duck without a nail. The frame was hand sawed and glued. It measures 30 feet long, 20 feet high and 15 feet wide. The eyes are taillights from a Model T Ford.

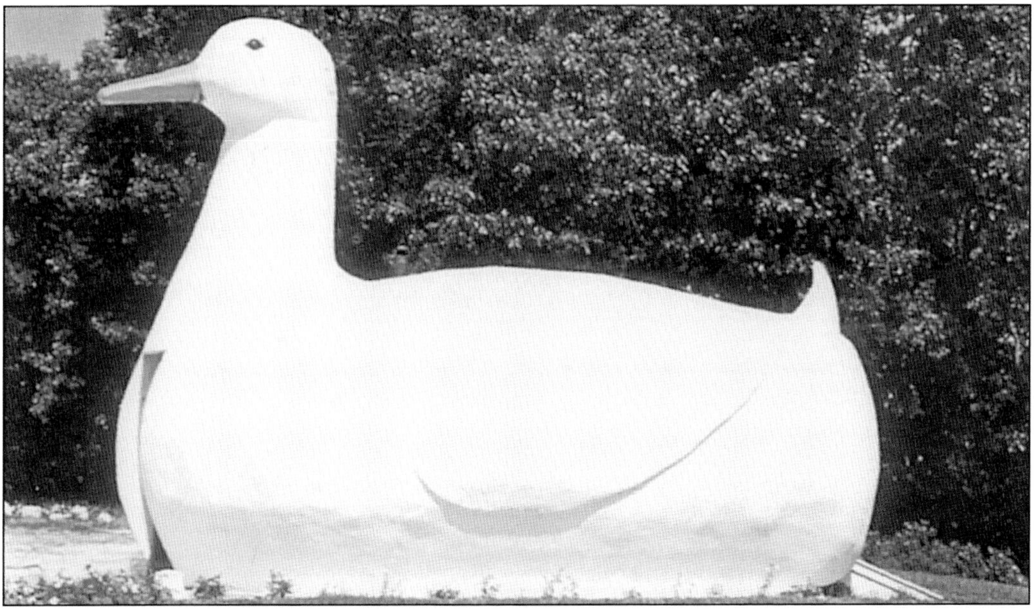

The Big Duck has been traveling. It was originally built on Main Street in Riverhead. In 1936, the Maurer's relocated to Route 24 in Flanders, taking the big duck along. Here it roosted for a half century. When the land was slated for development in 1987 the duck was donated to Suffolk County and moved to Sears Bellows County Park, where it was opened as a museum. Peking duck merchandise or "duck-a-bilia" are available in the Big Duck Store.

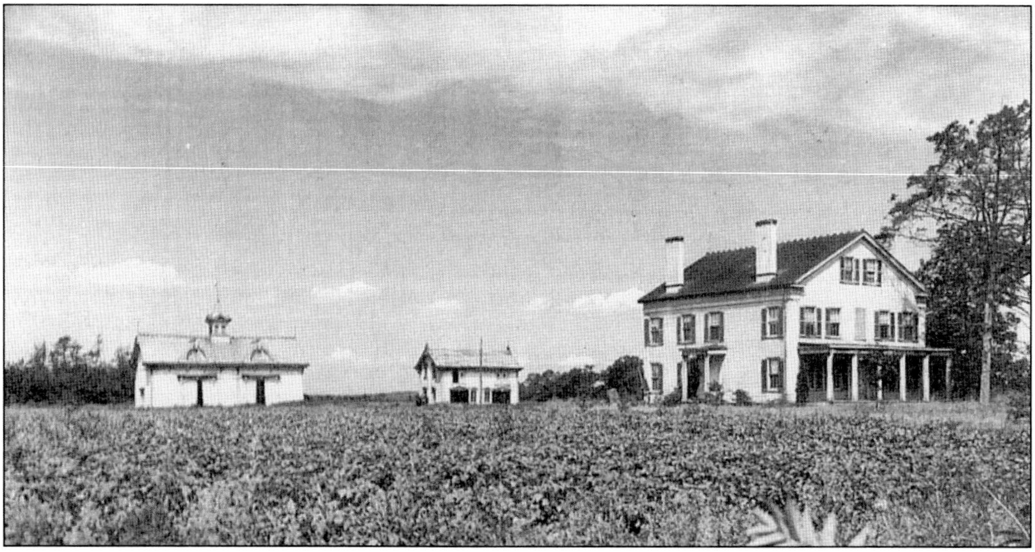

Potatoes, cauliflower, and melons were the principal products shipped from eastern Long Island to city markets early in the century. Long Island potatoes are famous the world over. They thrive well in Long Island's sandy soil and are sweetened by its nutrients. Before modern potato diggers were in use, hand pickers averaged 500 bags per day. Three to four acres are harvested per day using modern machinery.

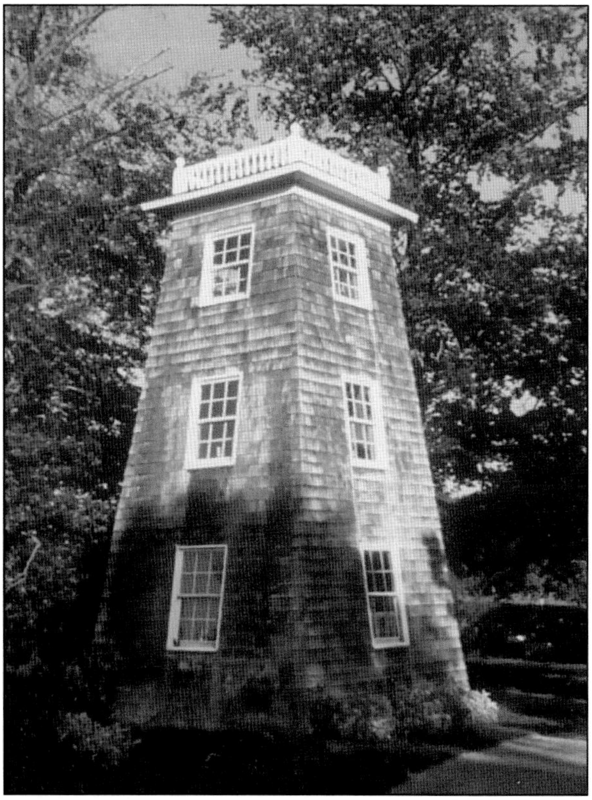

Center Moriches Mill was a farm mill whose sails have long since been removed. The art of converting wind into useful action lies with a specialist. The miller was a man of many talents. He was a weather forecaster, for it was necessary to anticipate the whims of the wind. He was an engineer and mechanic who maintained ingenious machinery. He was also a food expert, a craftsman, a local newscaster, and always a respected citizen.

Four

Wonders of Another Age

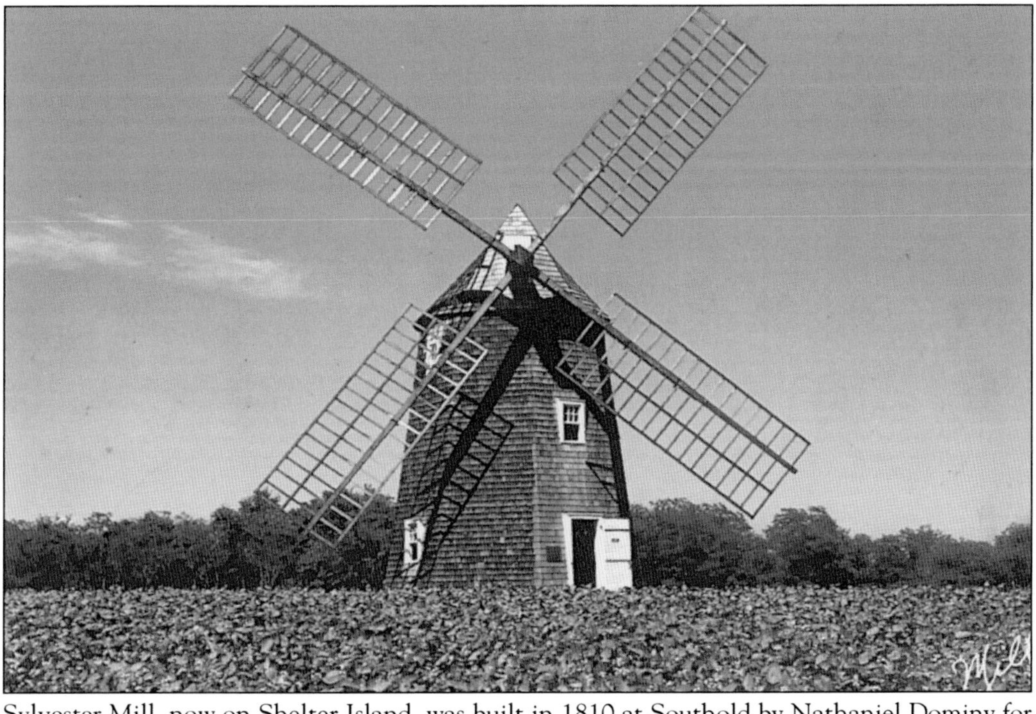

Sylvester Mill, now on Shelter Island, was built in 1810 at Southold by Nathaniel Dominy for the Southold Mill Co. It stood on Mill Hill at the western end of the village. Among Dominy's latest mills, it was equipped with separate girders to support the beams, which permitted greater headroom on the second level. A crown wheel and layshaft were placed under the ceiling. It is the only surviving mill built on the north fork of Long Island. Its sails were lost in a storm.

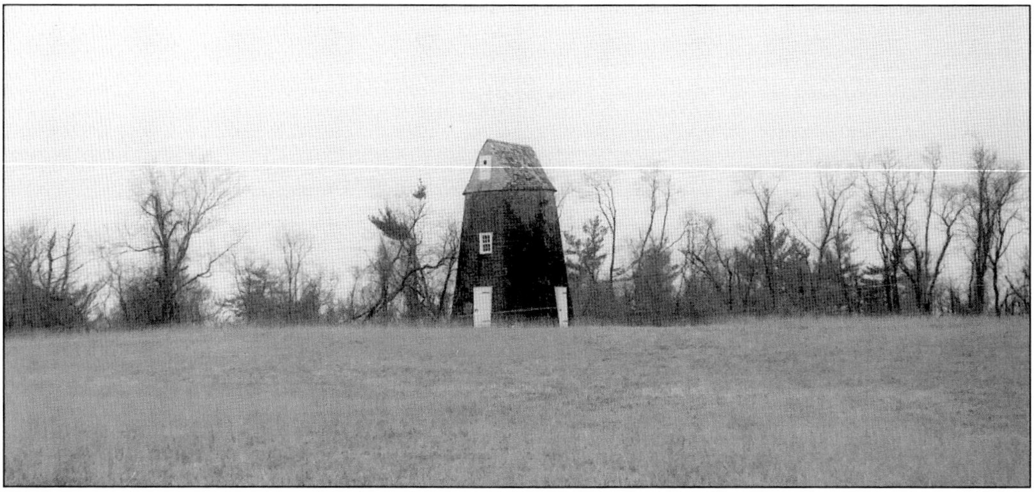

Sylvester's Mill on Shelter Island Mill contains one run of burr and rock stones. Two crown wheels are mounted on upright shafts above the spur wheel. In 1839, it was moved across Long Island Sound to Sylvester Manor on Shelter Island. Its owner during the war, Cornelia Horsford, had it renovated and worked in order to provide meal and flour to the inhabitants of Shelter Island during the food conservation period of World War I.

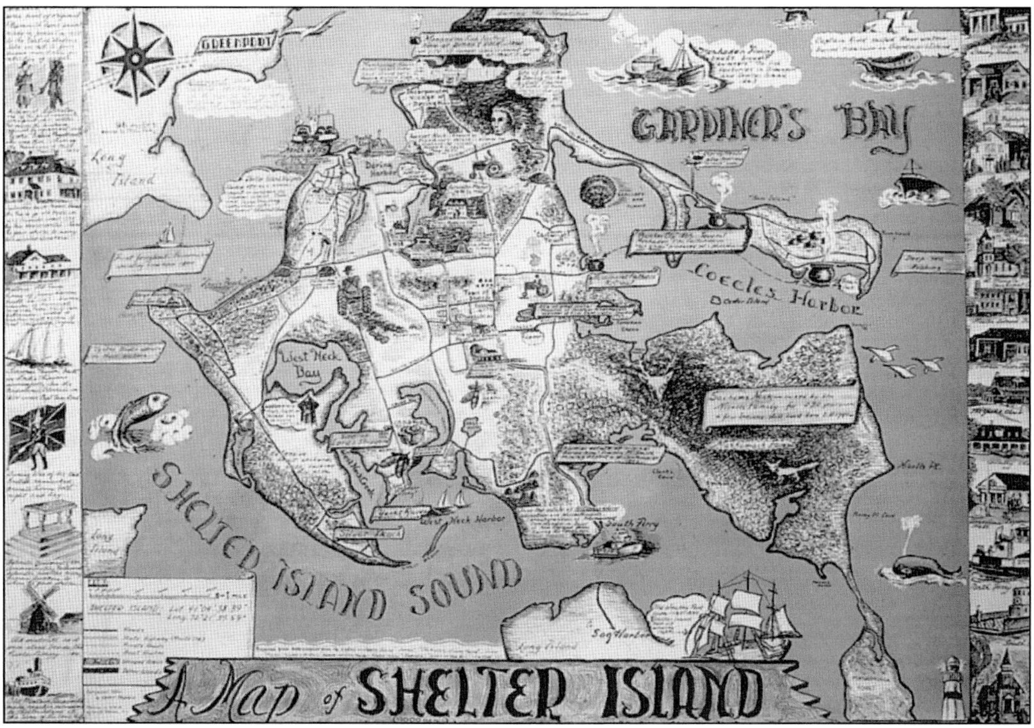

Shelter Island is an irregular shaped land situated between the two forks of eastern Long Island. In 1641, the owner, James Farrett, sold the land to four Royalists seeking shelter from European tyranny. One of the four, Nathan Sylvester, later bought the island and built a mansion. His descendants are still among its inhabitants. In 1656, during the religious persecutions on Long Island, the Quakers sought refuge on Shelter Island. It has become a renowned resort for New York City patrons.

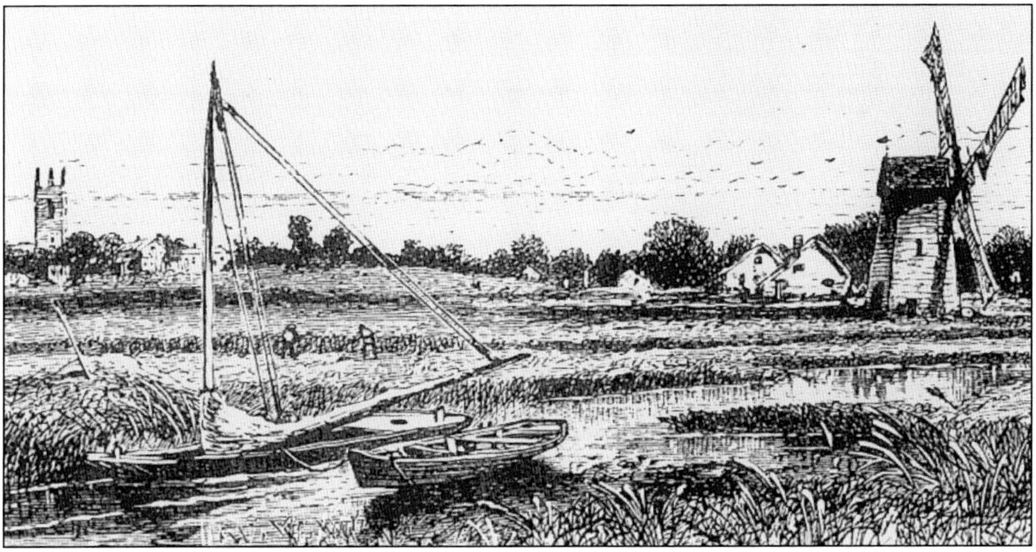

Nathaniel Dominy built Orient Windmill in 1810 for John Terry, a wheelwright in Orient. The mill had a gabled roof and internal winding machines. It was one of a very few built on the north fork of Long Island. The area, once called Oysterponds, is a peninsula. By 1873, the town boasted 500 inhabitants. It had a summer hotel and a wharf for the New York Steamer, used in transporting flour to New York City.

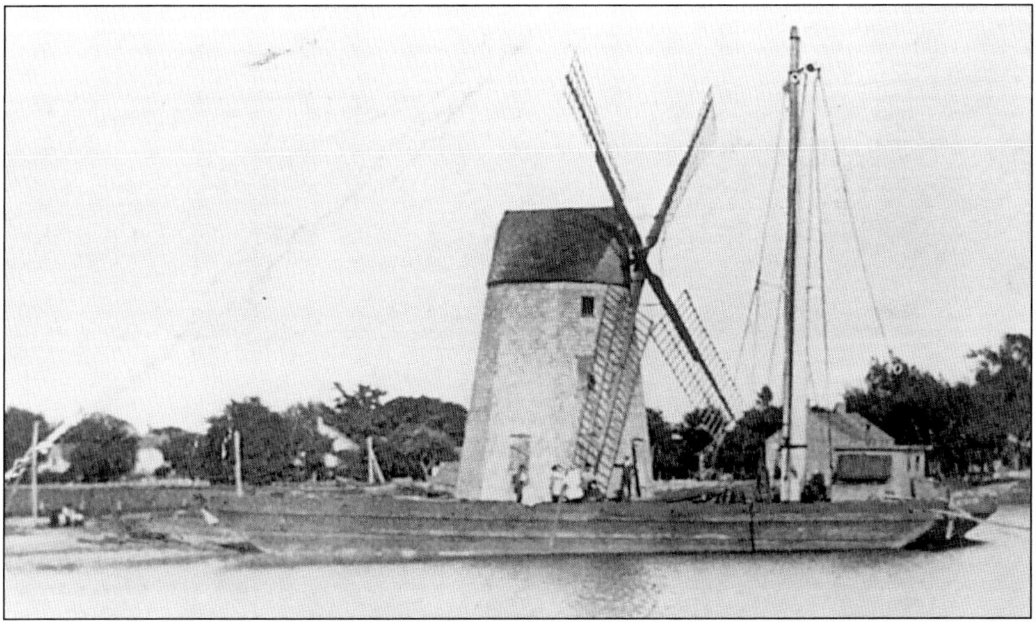

Orient Windmill was among the traveling mills. The windmill was just one century in operation when an iron cross was fixed to the wind shaft and the sail sockets were clamped and bolted for its journey westward. After a century of grinding on the north fork, it was placed on a barge and sailed to Glen Island. Many mills were relocated from place to place. It was easier and less expensive than constructing new ones.

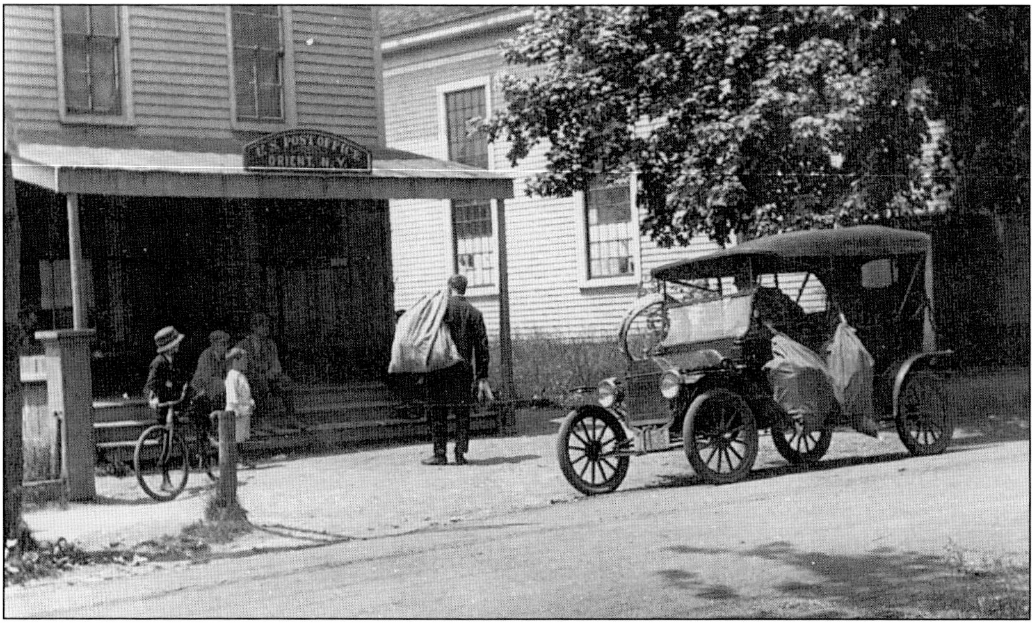

The mail car delivered mail to Orient Post Office in 1916. This service took place in the George Terry Country Store. Orient was called Oysterponds prior to 1836. The Oysterponds Historical Society Museum now encompasses the Historic District and core of Orient Village as it was in the 18th and 19th centuries. The village has retained its historic atmosphere.

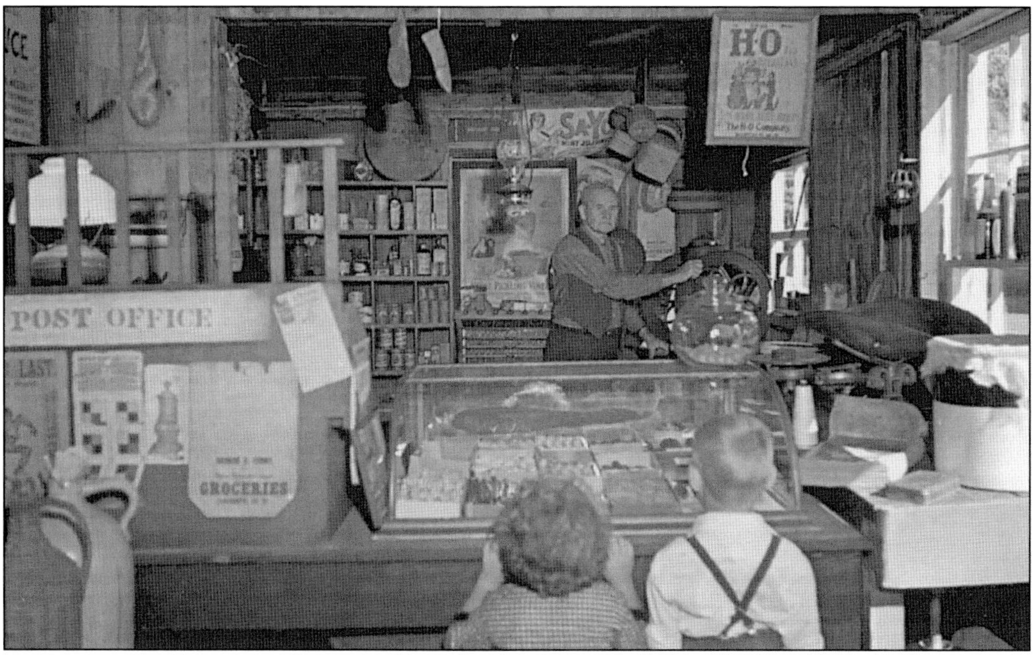

Children patronize George Terry's General Store where penny candy was a treat. Orient was settled in 1661. The earliest settlers built homes along what is now Main Street. A path led to the Wharf. The sheltered harbor had become busy with schooners carrying fish to New York City. Orient Point, five miles east, is the home of the New London Ferry, which transported passengers daily between Orient Point and New London, Connecticut.

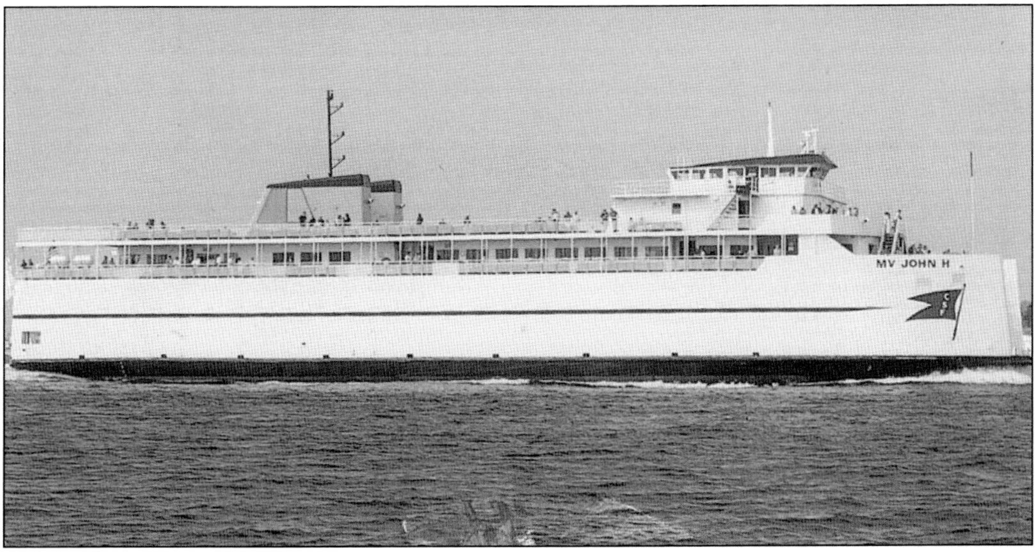

The M.V. *John H.* is a recent addition to the New London-Orient fleet. It was built exclusively for the Cross Sound Ferry in Panama City, Florida in 1989. The 240-foot luxurious liner is the largest in the fleet. It has the capacity to carry 120 vehicles and 1,000 passengers. Its amenities include comfortable indoor seating, children's play area, and on-board movies.

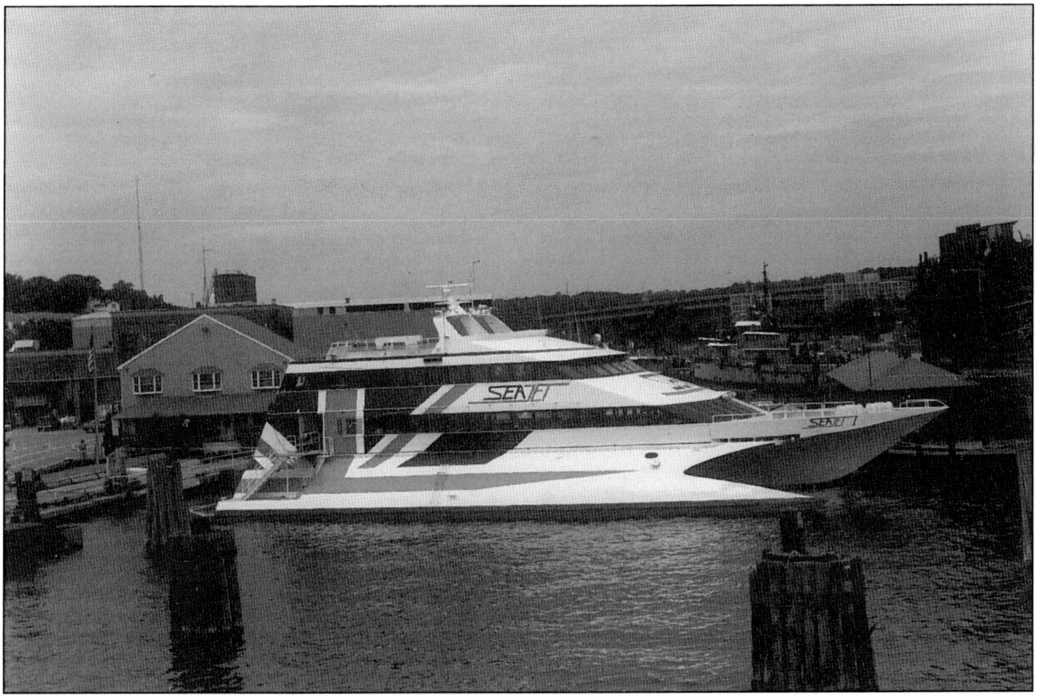

The *Sea Jet* is a newer and speedier ferry that makes the trip across Long Island Sound in half time. The 17-mile sail usually took nearly two hours. The *Sea Jet* makes it in 40 minutes. Cross Sound Ferry Services operate between Orient Point and New London, Connecticut. The *Sea Jet* has become a floating roadway or major transportation link between the Orient Point and New London, Connecticut, used by millions of travelers each year.

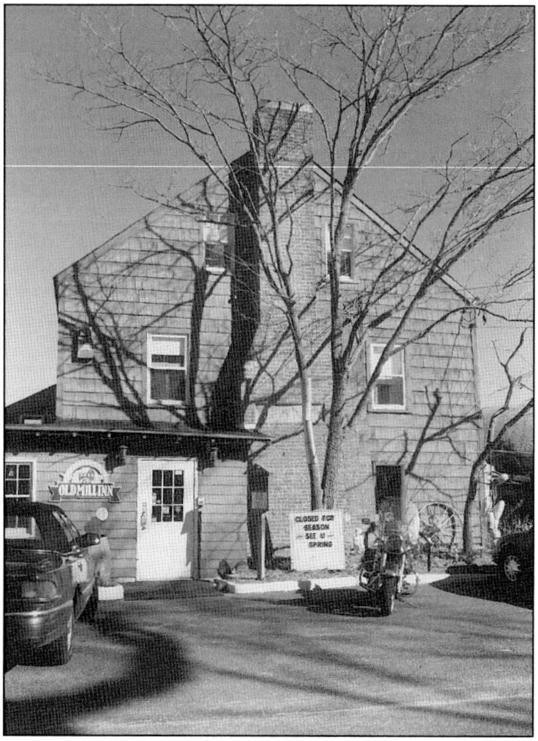

Sam Cox built Mattituck Tidal Mill on Mattituck Inlet in 1817. It was considered in the Library of Congress a rare engineering feat. The creek had to be dammed to obtain power for utilization and the schedule depended on the moon, tides, and water. In 1902, it was sold and evolved into a prominent restaurant known as Old Mill Inn.

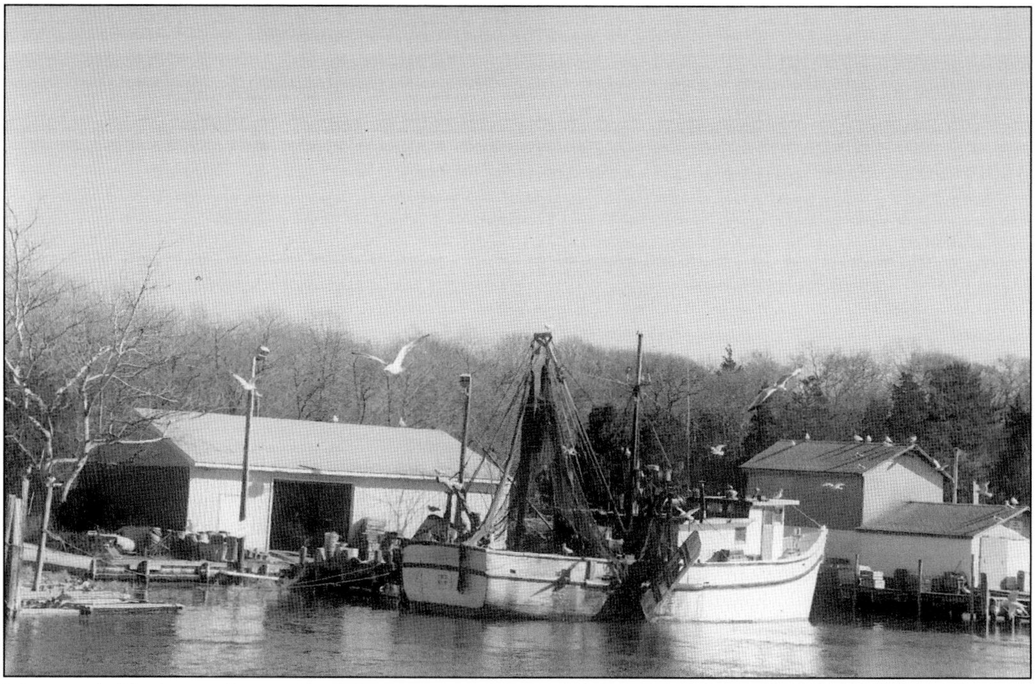

Mattituck Harbor is located on Great Peconic Bay. This is on the south side of the north fork. In cold and damp weather the miller lit a fire in his fireplace tucked somewhere in the building. His customers chatted while he worked. They gathered the latest news of the day, whispered the juiciest scandals, argued about politics, and debated over who caught the largest fish.

SAG HARBOR

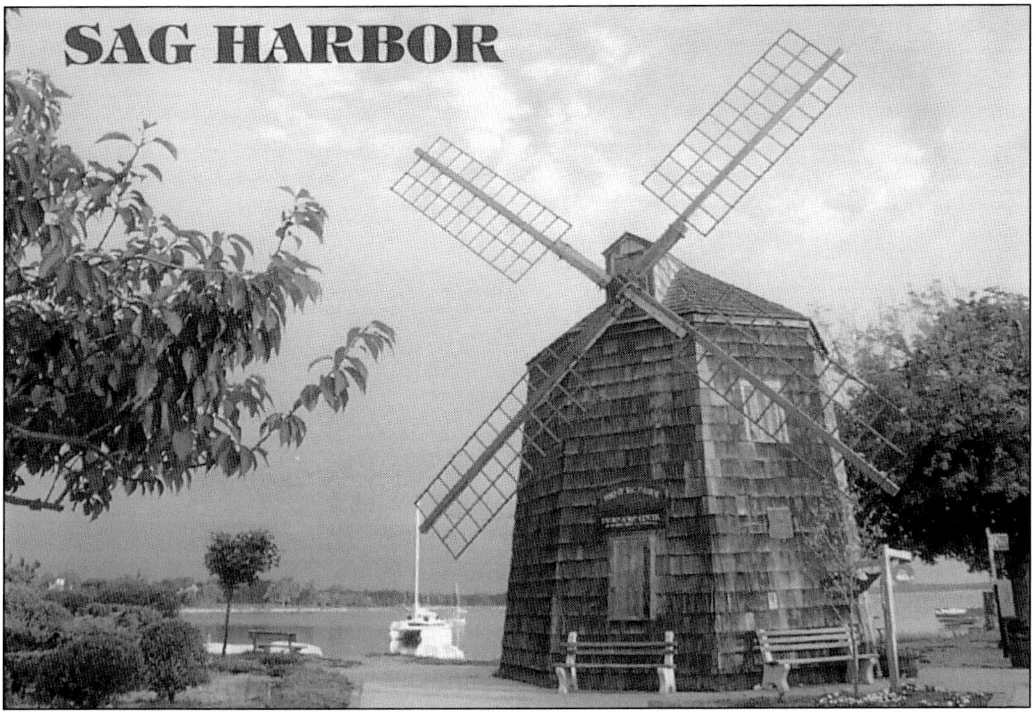

Sag Harbor was home to many windmills. The town was settled in 1707. The last workable mill was sold when the village became a prominent whaling port. A flag on the mill indicated a whale ship entering the harbor. In later years, a simulated windmill was constructed on Long Wharf. It serves as an information booth during the summer. Sag Harbor is home to the ferry that plies the bay to Shelter Island.

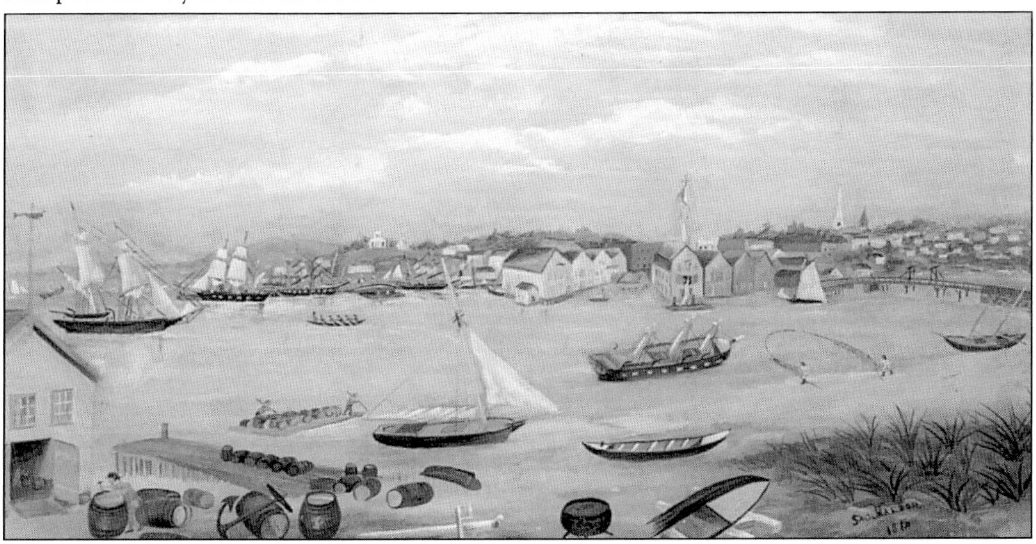

Sag Harbor gained distinction in 1789 when President Washington chartered it as a port of entry. In the early 1800s, it became the center of the whale industry, and by 1847, 275 whale ships were anchored in port bringing $7 million in whale-related products. The first customhouse in the country, the first newspaper on Long Island, and the first appointed postmaster in New York State all became part of Sag Harbor's heritage.

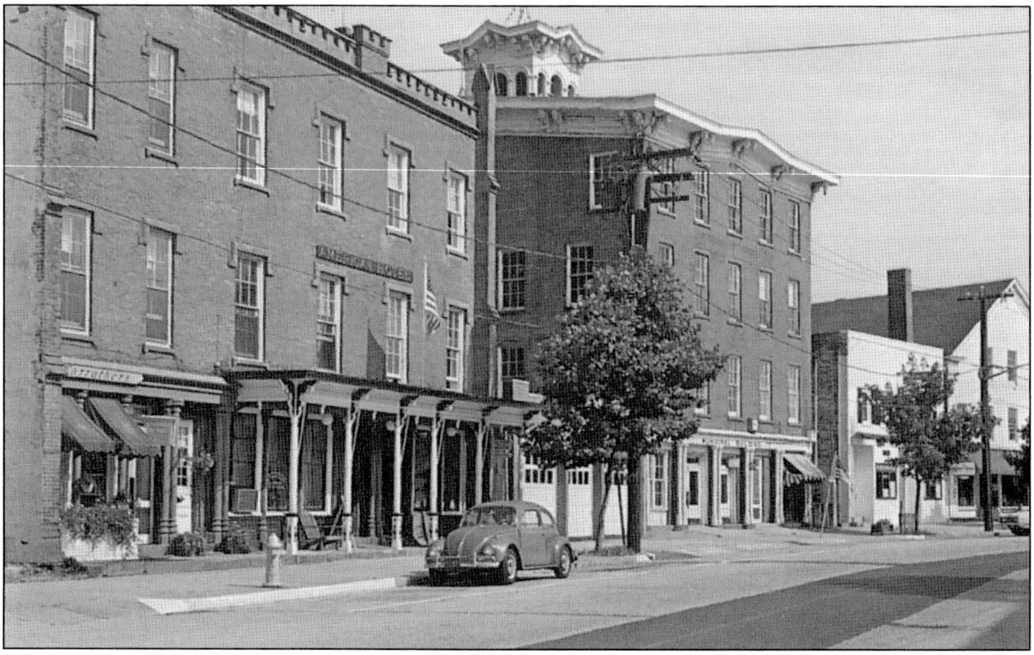

The American Hotel and Municipal building are replicas of whaling days in Sag Harbor. The hotel has been a landmark since 1846 and represents a rare example of rural hotels that flourished in the 19th century. It was here that the whalers refreshed themselves after trips at sea. It is also where James Feinmore Cooper wrote his famous *Sea Tales*.

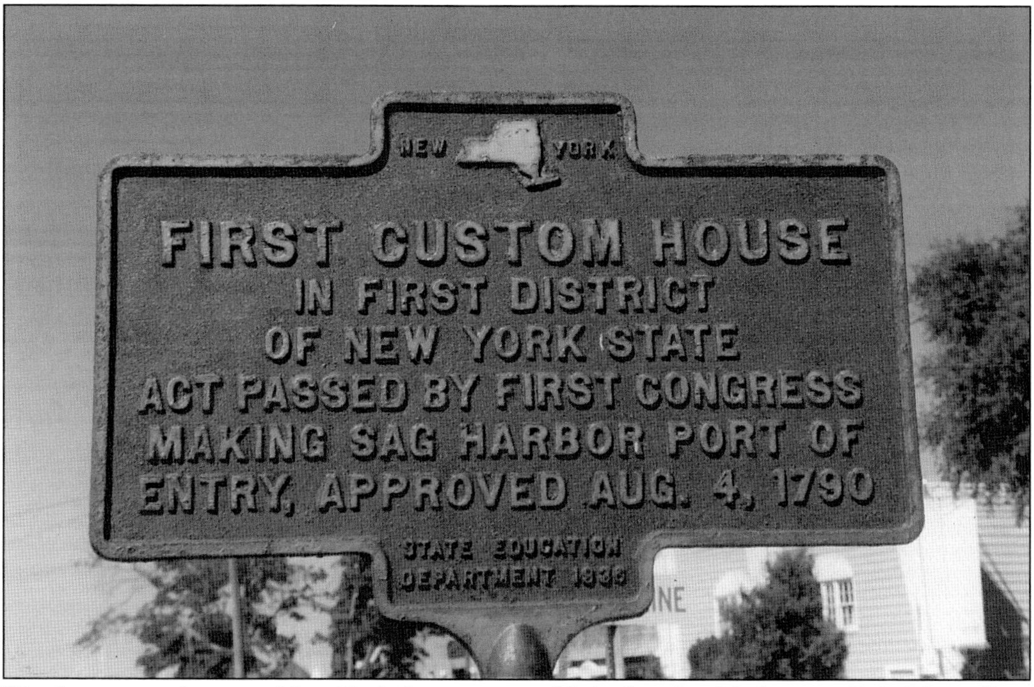

The first customhouse in New York State was established in Sag Harbor in 1790. The village attained this distinction just before it reached the status of the fourth largest whaling port in the world.

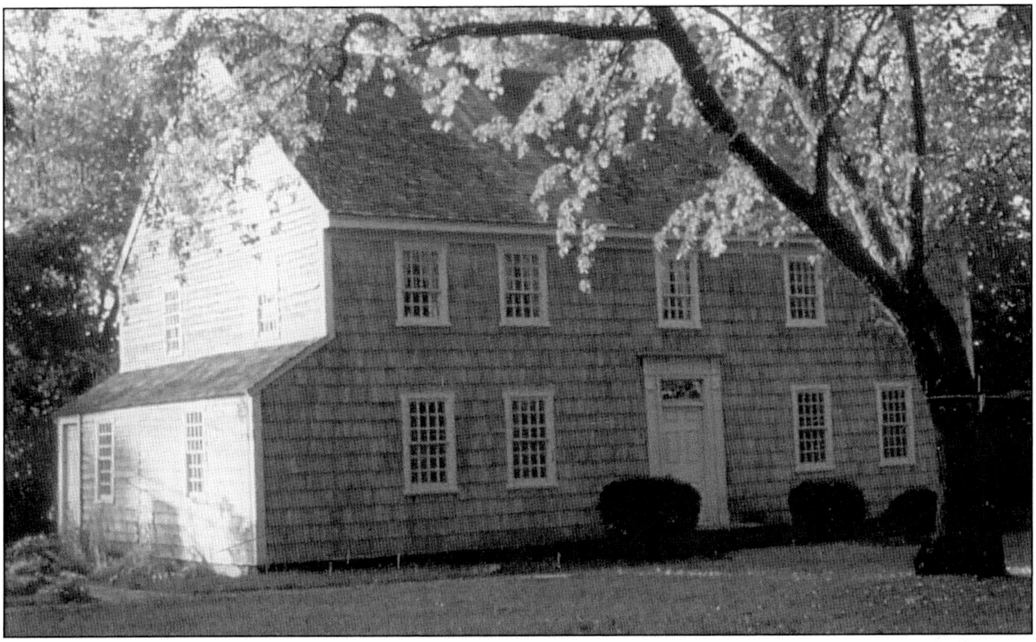

Sag Harbor's old customhouse was established in 1789 on Union Street. It was moved to its present location through the generosity of Charles Edison, son of the famous inventor. All ships manifests were inspected and cleared, duties and dues collected, and United States mail delivered in this building. The first collector of the port was Henry Packer Dering, who was also designated as the local postmaster.

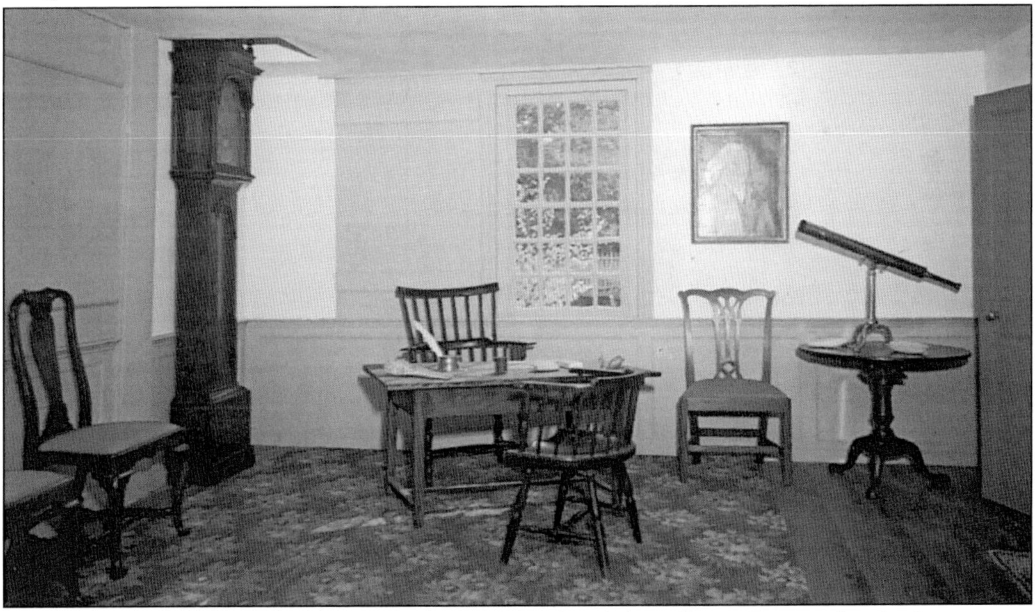

The customhouse in Sag Harbor reveals the lifestyle of the affluent Dering family. It contains furnishings fashioned by east end craftsmen. Henry Packer Dering was customs master for three decades. He served the young republic as chief federal agent during the War of 1812. His duty was to meet the trading vessels and whaling ships that entered local waters. In the Federal Customs House he not only conducted customs affairs but raised nine children.

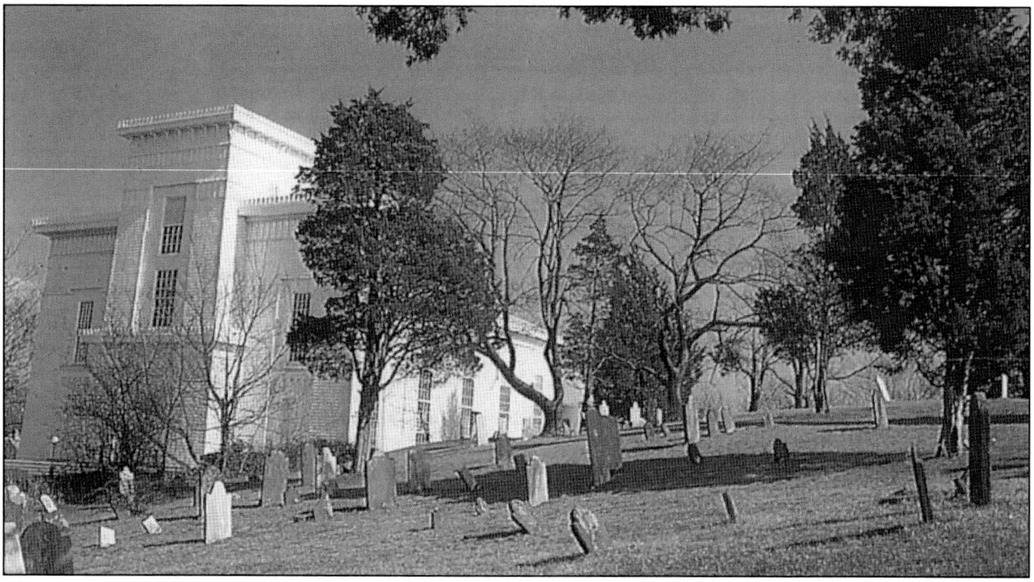

The Presbyterian church at Sag Harbor was erected in 1842. The architecture is in keeping with the historical nature of the village. Designed by Minard Lafever, a noted architect of the day, it became the most renowned example of Egyptian Revival style in the United States. This architecture is symbolized in the pylons that taper upward and the central base that supported the steeple. The church has made many cultural contributions to the area.

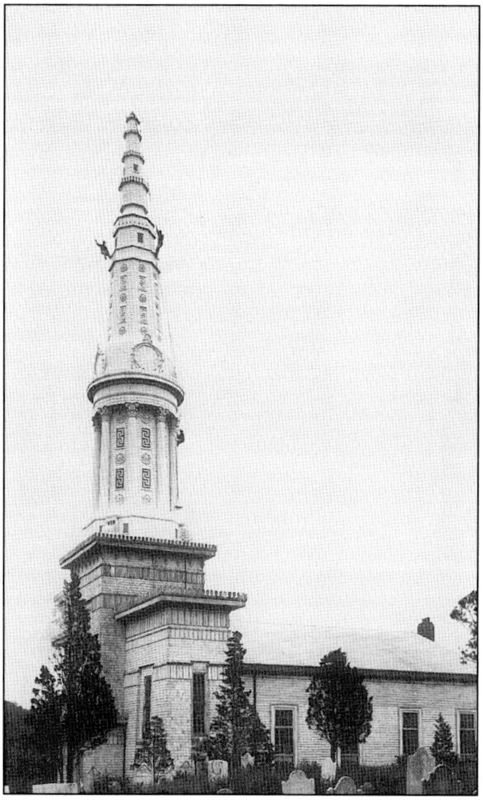

The steeple of Old Whalers Church towered 185 feet into the sky. It created a focal point for incoming ships. The spire tapered upward to resemble a lighthouse. The lower section contained the bell and the middle section featured clocks facing in four directions. The slender spire reached to a weathervane. The violent hurricane of September 1938 toppled the stately landmark to the ground. The steeple lifted from the building and landed beside the church, which was unharmed.

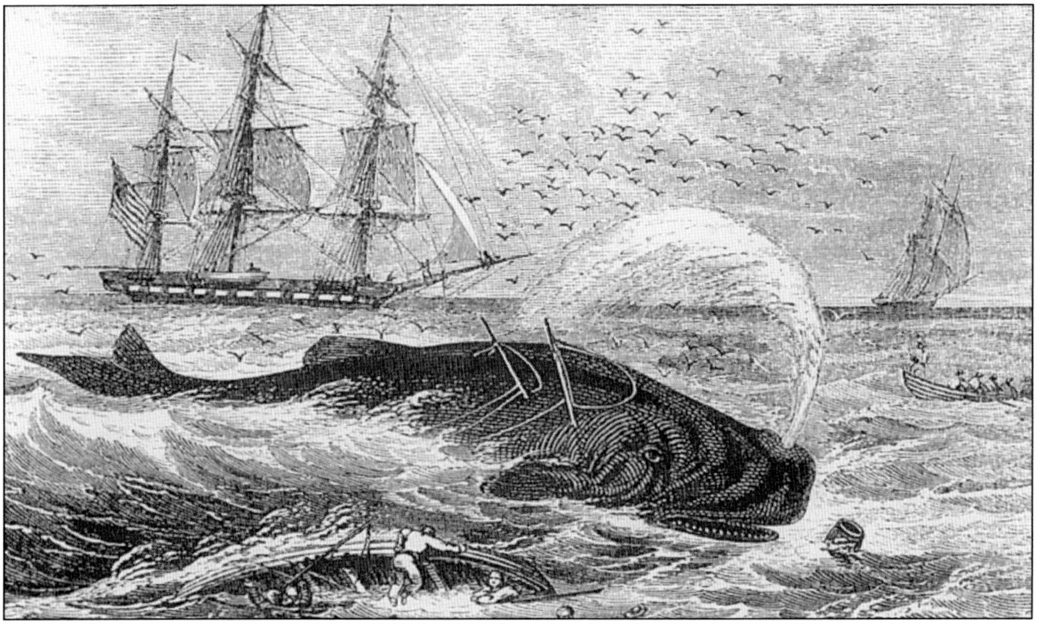

Whales were originally captured in Long Island waters. Local whales became scarce and sailors were required to travel farther. Sperm whales inhabit tropical waters and this quest took the whaling ships on extensive voyages. The whale contained valuable oil used in making candles. When a whale was discovered the sailors used a small whaleboat to harpoon the whale.

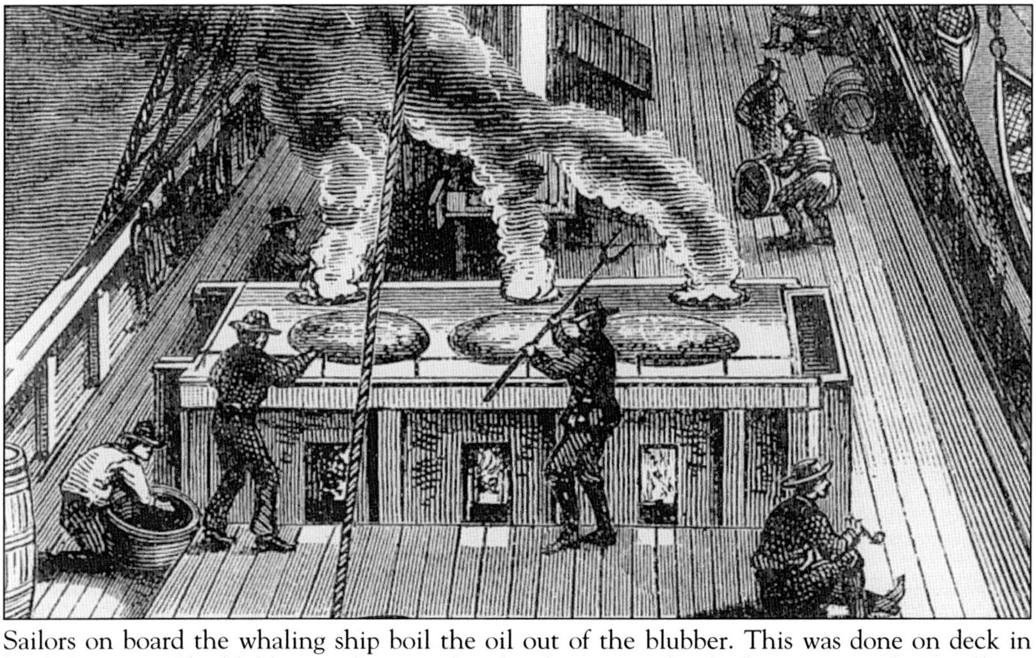

Sailors on board the whaling ship boil the oil out of the blubber. This was done on deck in brick furnaces. The blubber, or fat, was placed in huge pots that had a capacity of 400 gallons. The precious oil was carefully stored. Whalebones and ivory were brought home. Whaling ships were often on four-year voyages. The process of boiling blubber was called "trying out."

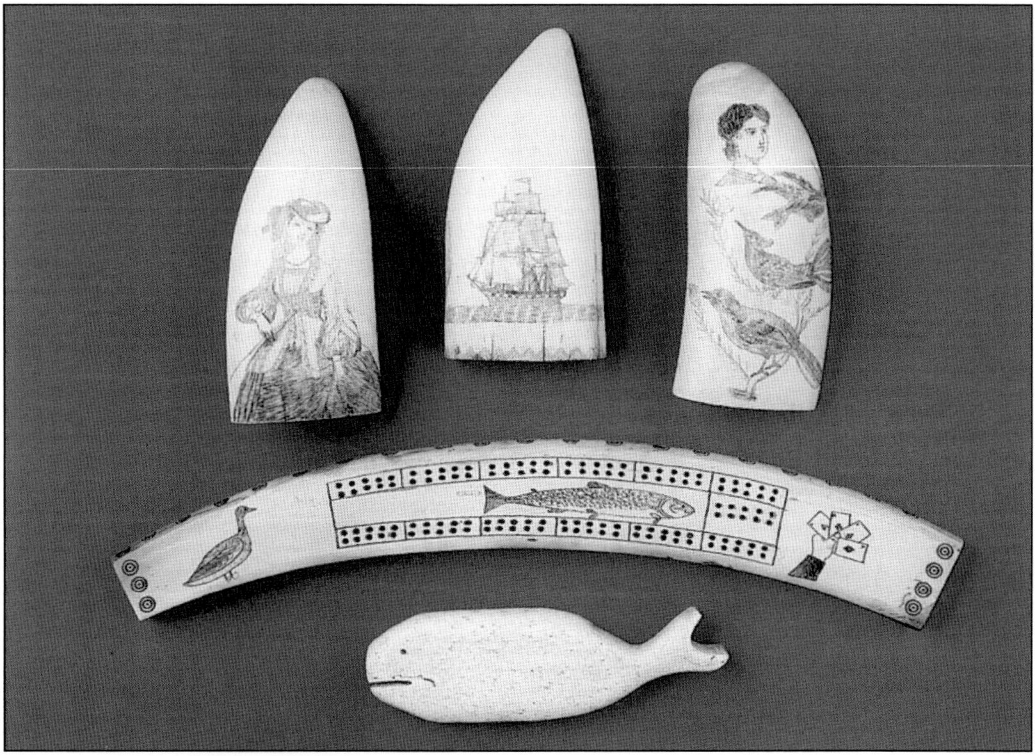

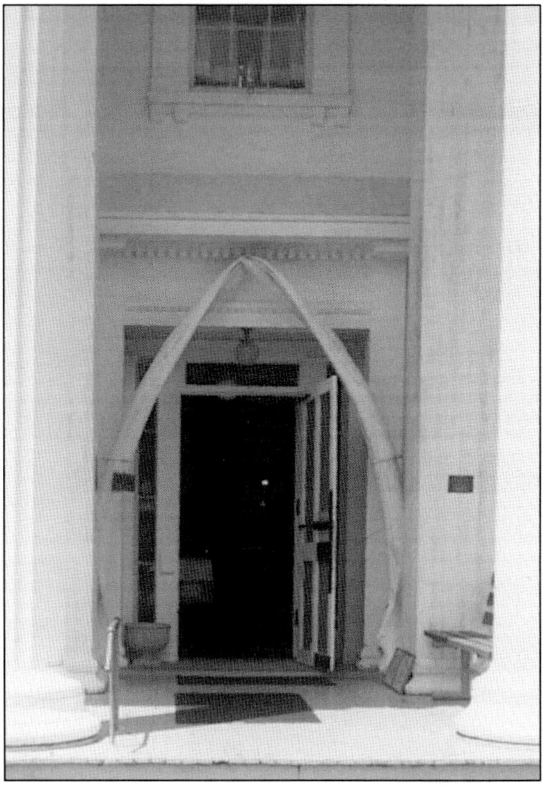

Scrimshaw art was a common pastime among the sailors on the whaling ships. They spent years at sea capturing whales and became very skillful at sculpturing scrimshaw, which was brought home and can be viewed in Sag Harbor. The photo depicts engraved scrimshaw teeth, a cribbage board and whale, whale bone, and ink. Whalers also crafted ivory at sea.

Sag Harbor Museum, opened in 1936, contains many artifacts of whaling days. A collection of scrimshaws is among the treasures of antiquity. Most significant is the authentic jawbone of a whale over the entrance. When one enters the museum one has just gone through the jawbones of a whale.

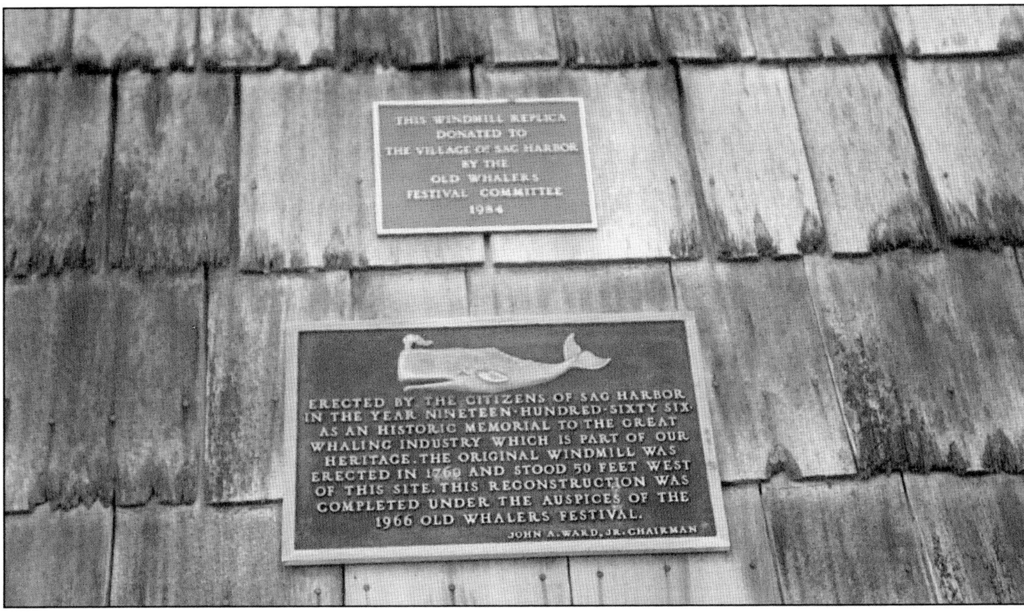

The shingles of Sag Harbor replica windmill hold the plaque that is a reminder of Sag Harbor's historic past. Sag Harbor was the most significant whaling port in the country. Large square-rigged vessels sailed from Sag Harbor on extended missions and usually returned with a shipload of whale products or "whale-a-bilia."

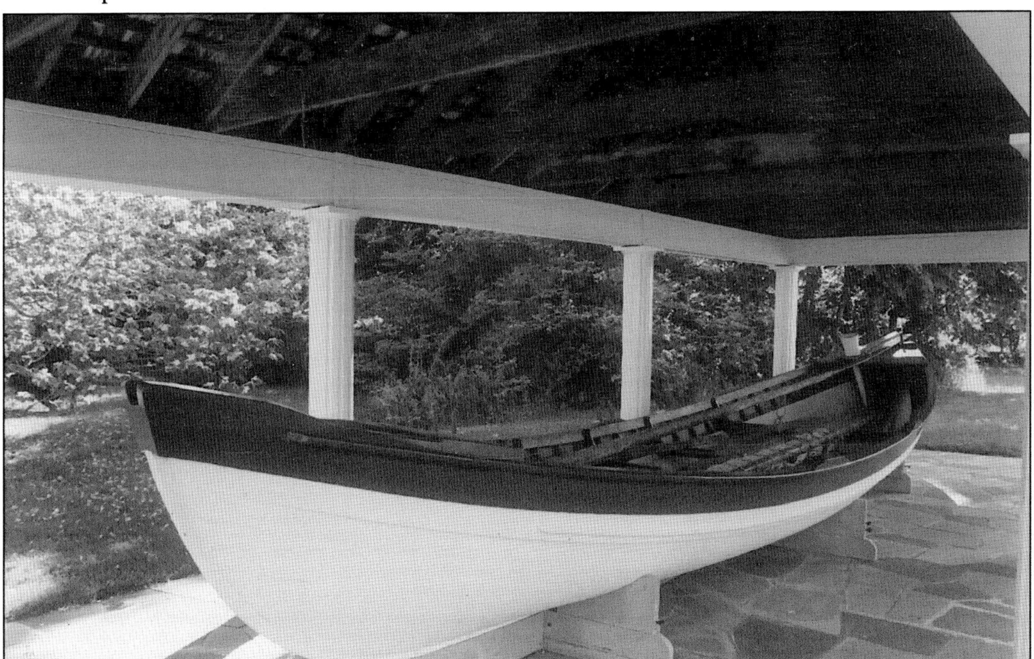

Small whaleboats were a necessity on whaling expeditions. The settlers learned from Native Americans the art of whaling. They originally used dugouts hewed of huge trees. In the early 1800s, there was a scarcity of whales in Long Island waters and off-shore whaling became prominent. Iron harpoons and light whaleboats made of native cedar were taken on the whaling expeditions. This boat is displayed at Sag Harbor.

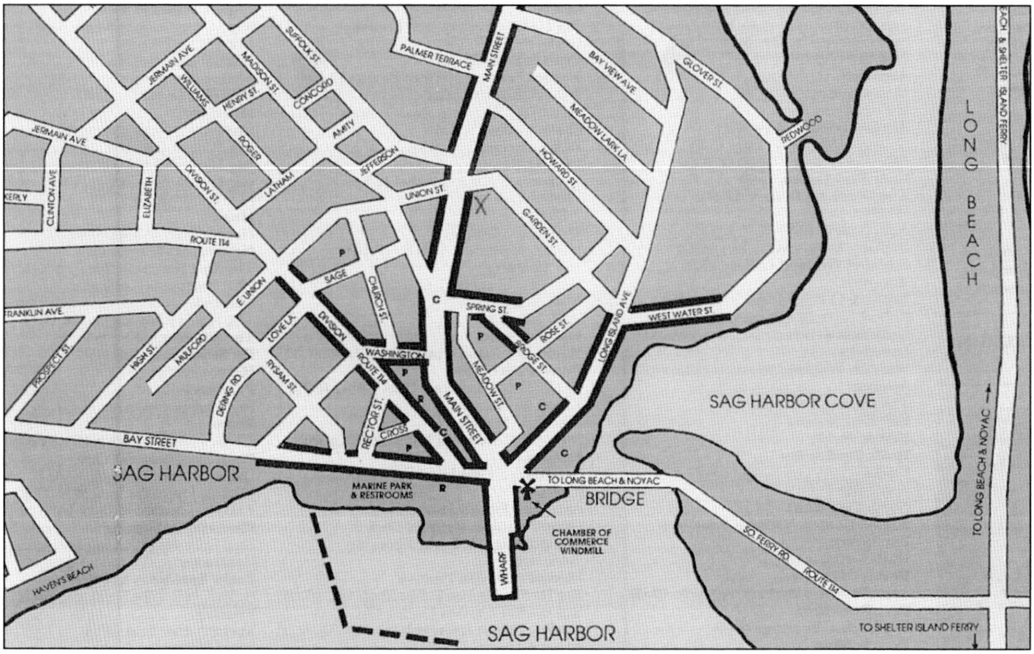

The Map of Sag Harbor indicates its proximity to the sea. It was from Long Wharf that the whaling ships departed on their expeditions. They returned with their "whale-a-bilia" to the same pier. Whaling declined with the discovery of oil and many of the whaling ships sailed to the gold rush in California in 1849. The X marks the site of the customhouse.

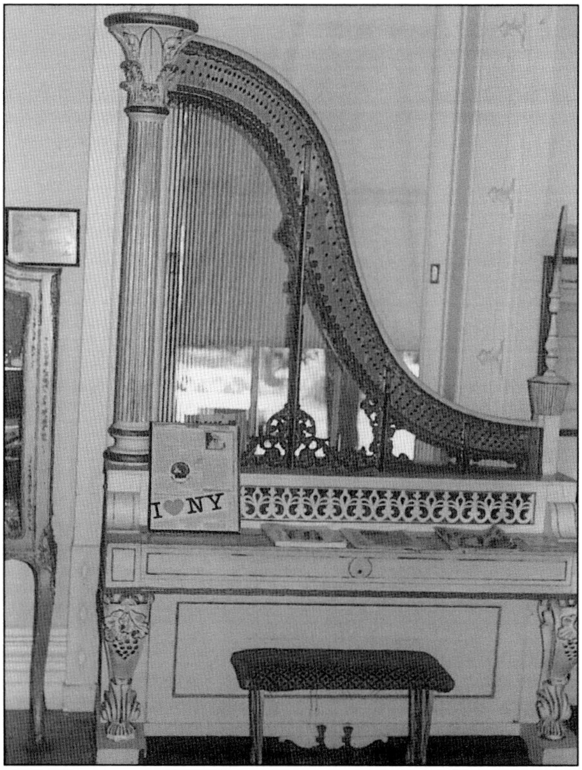

This is a Kuhn Giraffe Piano built 1860–80. It is also known as a clavicytherium. The upright harpsichord, with a vertical soundboard, is part of the Whaling Museum collection. Minimized space and direct sound projection involves use of springs rather than gravity to return jacks to their proper position. The clavicytherium is closed so the keyboard is not visible.

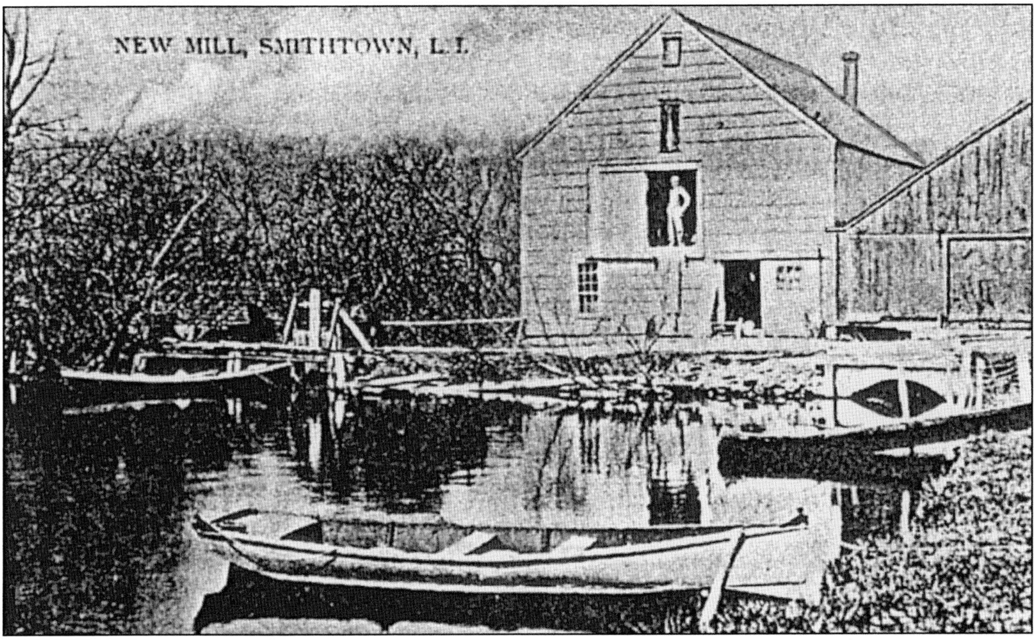

The Blydenburgh Milling Center had its beginnings in 1715. Three farmers in Smithtown agreed to build a complex of water wheel mills at the junction of their farms. They all bordered the Nissequogue River. Dams were necessary to stem the tide and provide power for the mills. The first mill was constructed in 1715 at the head of the river and is now part of a private residence. Many water wheel mills were constructed on tidal inlets before the Industrial Revolution.

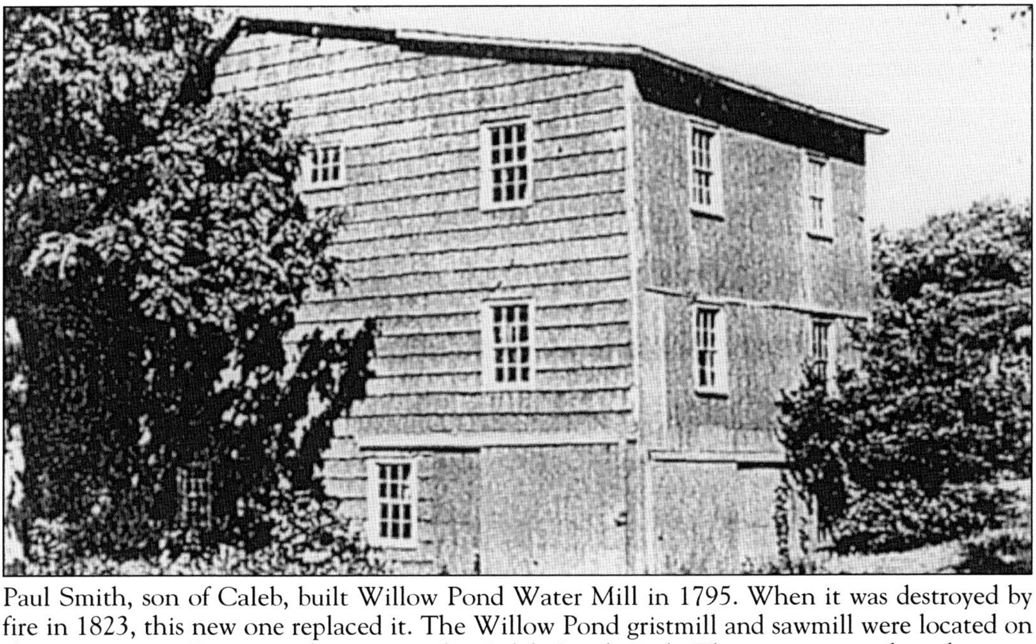

Paul Smith, son of Caleb, built Willow Pond Water Mill in 1795. When it was destroyed by fire in 1823, this new one replaced it. The Willow Pond gristmill and sawmill were located on the Nissequogue River in what is now the Caleb Smith Park. The area surrounding the river was becoming industrialized with the latest technology of the 18th and 19th centuries. It was known as The Mill Center.

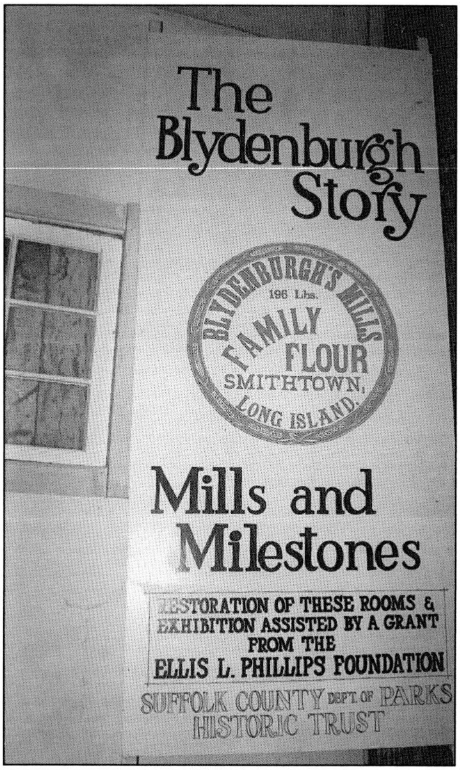

The Blydenburgh Mill was constructed in 1798 as an overshot water wheel mill. In 1870, it was enlarged to accommodate the new machinery of roller milling. This alteration is detectable in the form of varied-size shingles on the building. Roller milling represented a technological advance over the water wheel system. This is one of very few roller mills on Long Island. The mill was a vibrant part of economic and social life.

The original Blydenburgh Mill complex of Smithtown consisted of sawmills and woolen mills. The woolen mill was originally the First Presbyterian church, which was located at the junction of Moriches and Nissequogue River Roads. Constructed prior to 1750, it was moved from The Branch to Blydenburgh Park. It included a farmhouse complex in the Victorian period. These mills illustrate much of the farm-to-mill-to-market cycle of pre-industrial American economy.

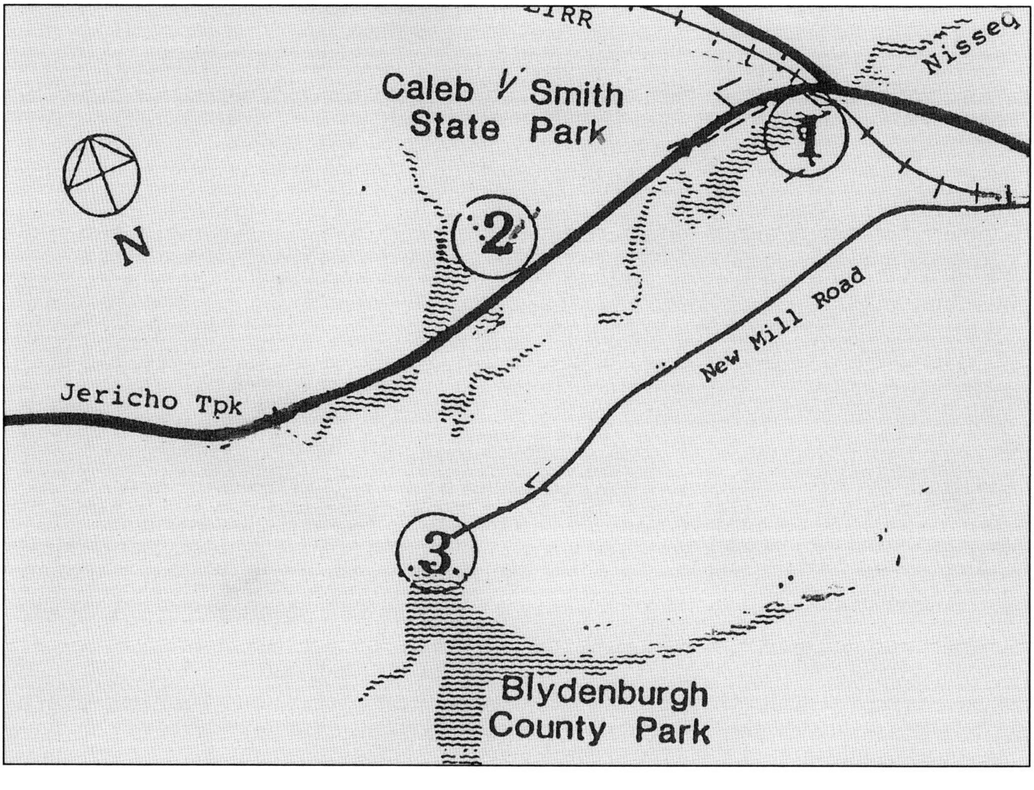

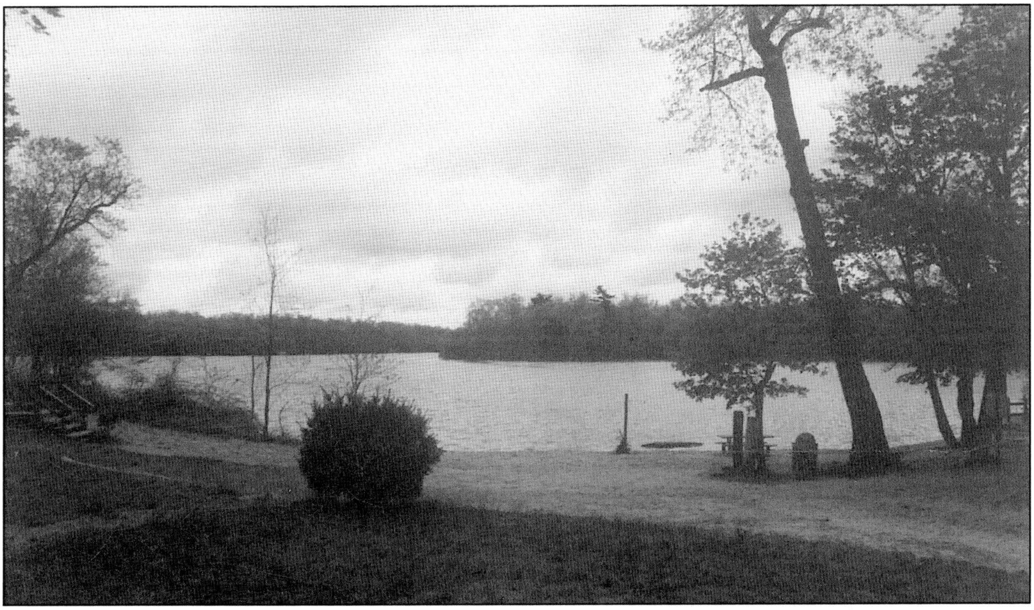

Stump Pond was necessary to barricade water before the saw and grist mills could operate. The farmers, Caleb Smith, his brother Joshua, and Isaac Blydenburgh flowed water over acres of forest land where only the stumps of trees remained. These stumps have been under water for nearly two centuries. It was at Stump Pond that Blydenburgh Mill was established.

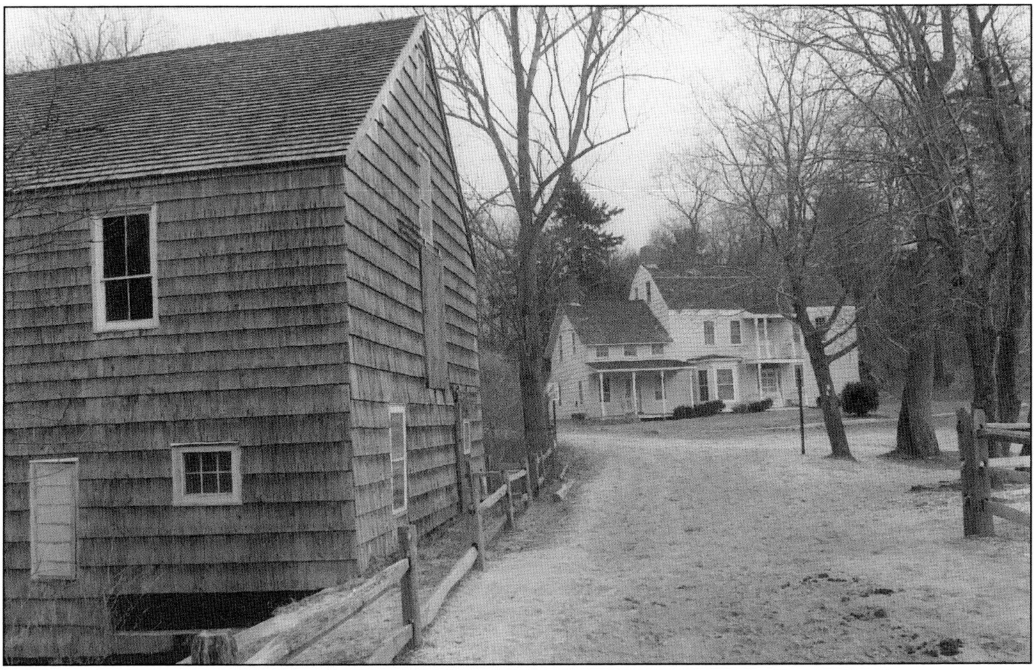

Isaac Smith built the Miller's house in 1803. It was positioned on Bushy Hill adjacent to the mill. It commands a view of New Mill and the 580-acre Stump Pond. The house surpasses ordinary country dwellings. It embodies a rare plaster technique and unusual porticos at the entrance. The Blydenburg Mill, in the foreground, is often called the New Mill to distinguish it from two earlier 18th-century mills on the Nissequogue River.

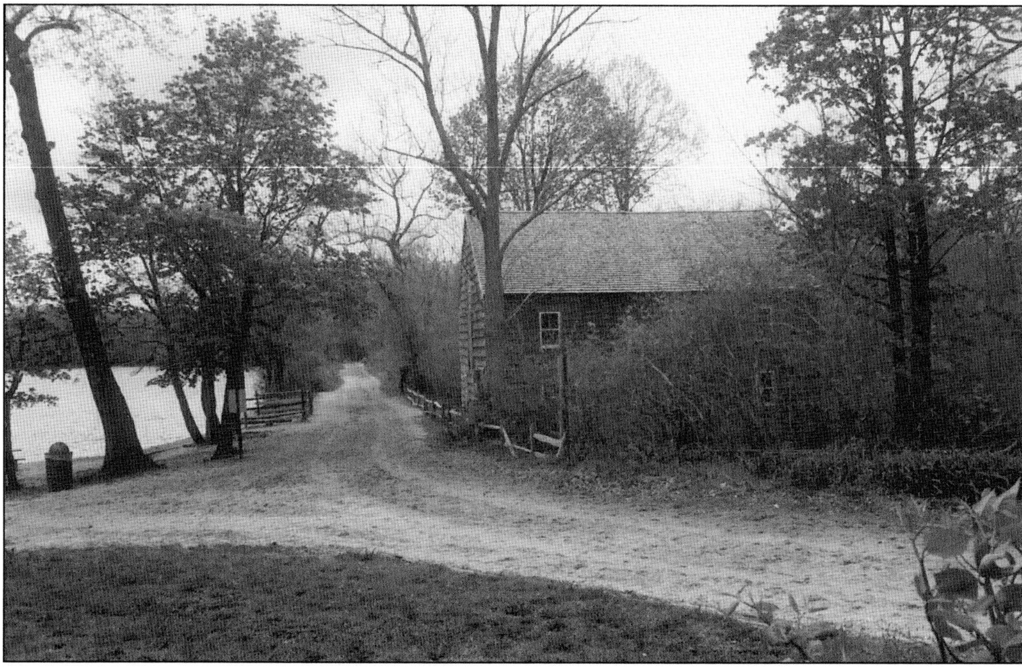

A rustic path surrounds Stump Pond as one looks south from the farmhouse. In Caleb Smith State Park and Blydenburgh State Park today, one is hardly aware of their contribution to industrial pursuits and technology of the 18th century. Smithtown enjoyed a thriving business on the Nissequogue River.

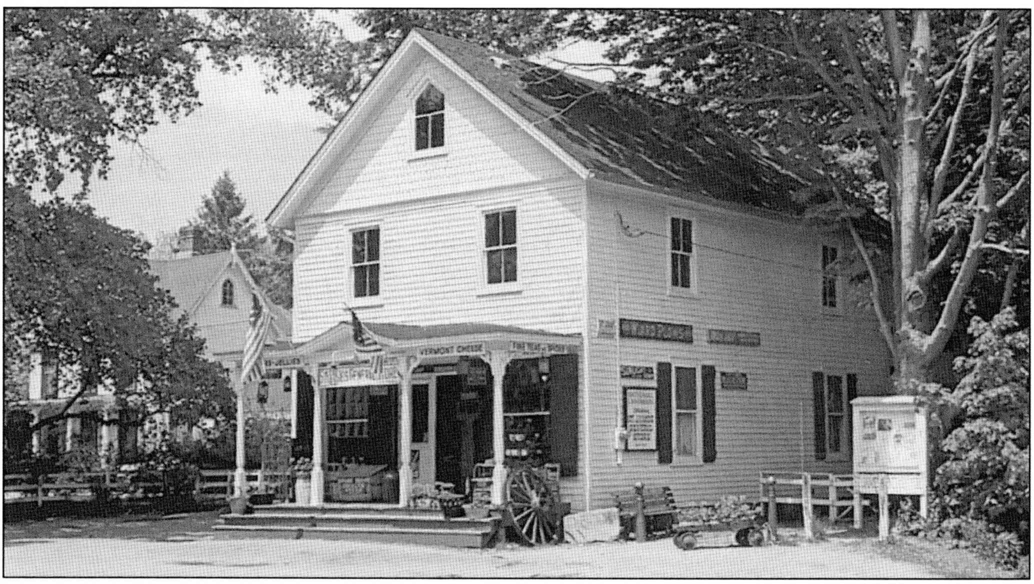

St. James General Store has been in continuous operation since 1859. Ebenizer Smith, a descendant of the legendary bull rider, built it. It was a central meeting place where people waited for mail, caught up on gossip, and kept in touch with the world. There were parties and dances held in the upstairs room. The first telephone in St. James was installed in the store, which later became a community center. It is the most authentic general store in the country and has never been modernized.

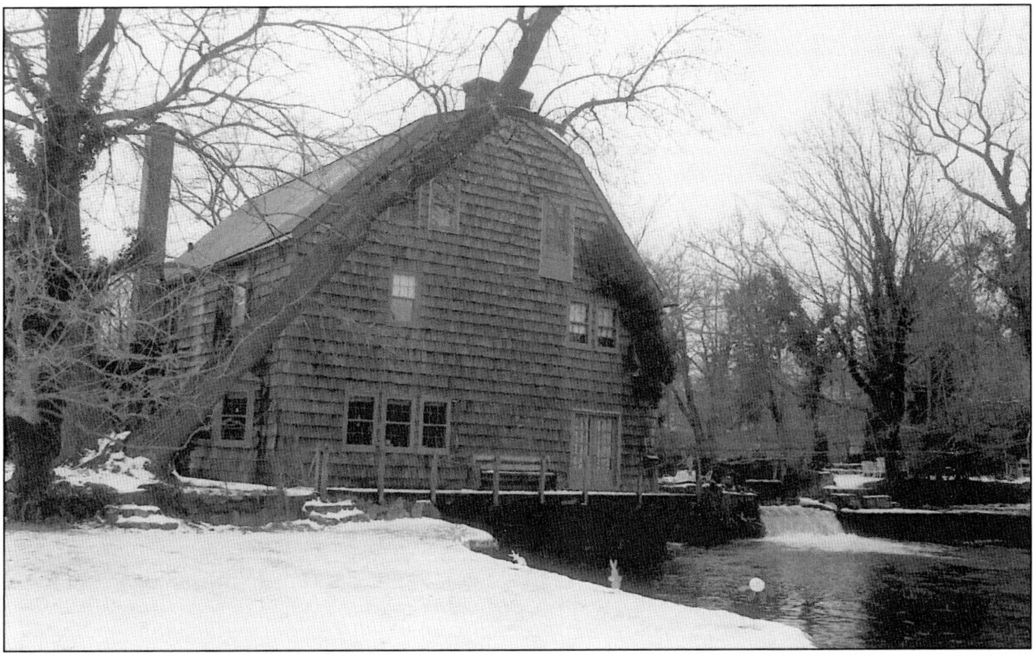

Phillips Mill in Smithtown dates back to 1720. Amos Willetts and Daniel Bates built it for George Phillips, son of the first Setauket Minister. It was situated near the head of the Nissequogue River, now "Head of the Harbor." It was a three-story mill powered by a horizontal waterwheel. It closed in 1909 and has been a private residence since 1927. Head of the Harbor is a little hamlet nestled between Stony Brook and Smithtown.

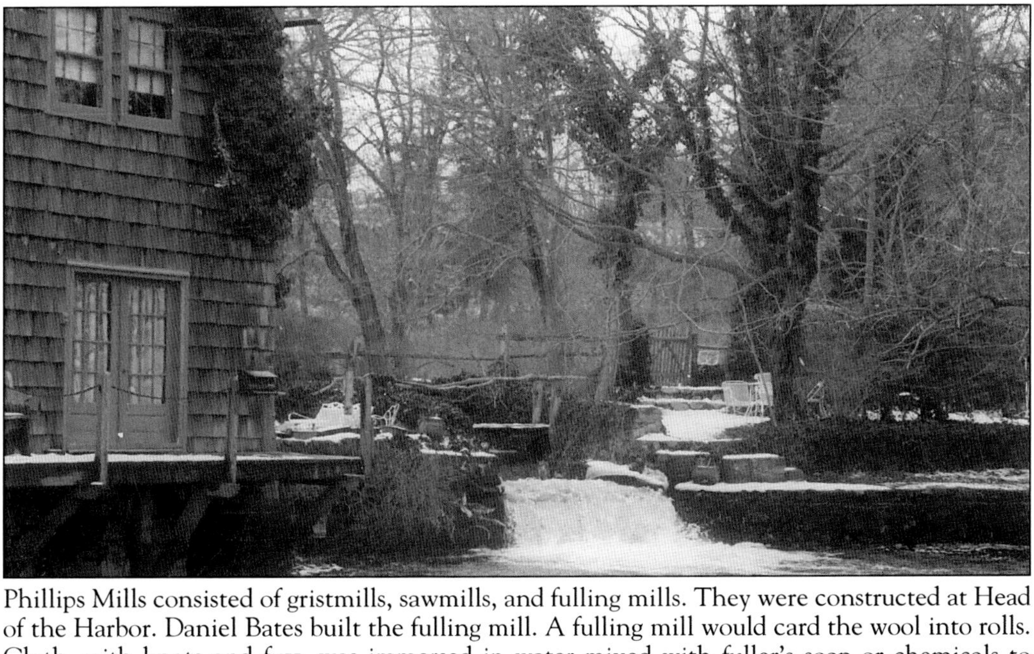

Phillips Mills consisted of gristmills, sawmills, and fulling mills. They were constructed at Head of the Harbor. Daniel Bates built the fulling mill. A fulling mill would card the wool into rolls. Cloth, with knots and fuzz, was immersed in water mixed with fuller's soap or chemicals to remove impurities. The cloth was pounded by water-driven mallets to beat the woven threads into closer form. It was then stretched on tender hooks to dry and prevent undue shrinkage.

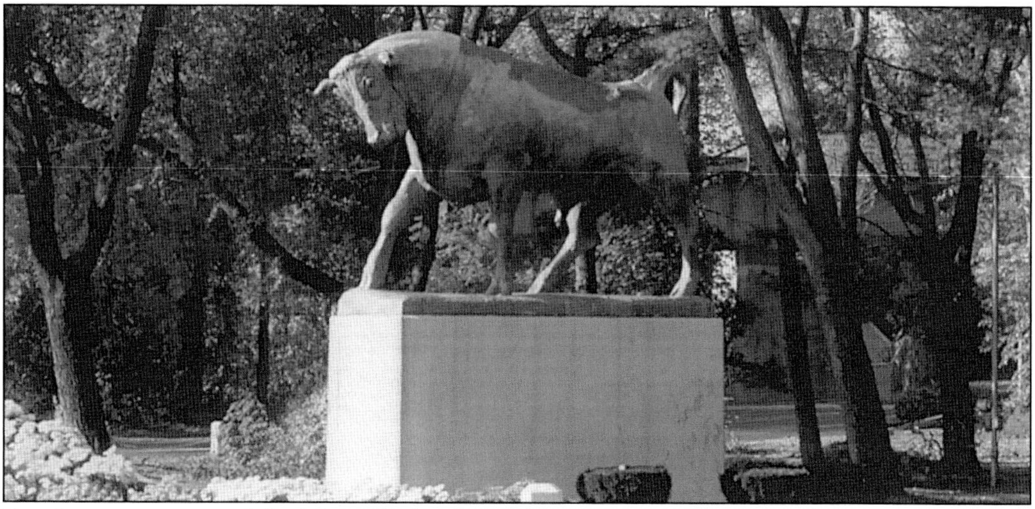

Smithtown was named for Major Richard Smith in 1665. Legend has it he agreed to buy all the land he could ride around on the back of a bull from sunrise to sunset. Accomplishing this feat around his desired parcel of real estate may have been in accord with tribal customs of the Nissequogue Indians, who originally owned the land. The bronze bull "Whisper" stands at the junction of Routes 25 and 25A, and was designed by Charles Ramsey of Old Westbury.

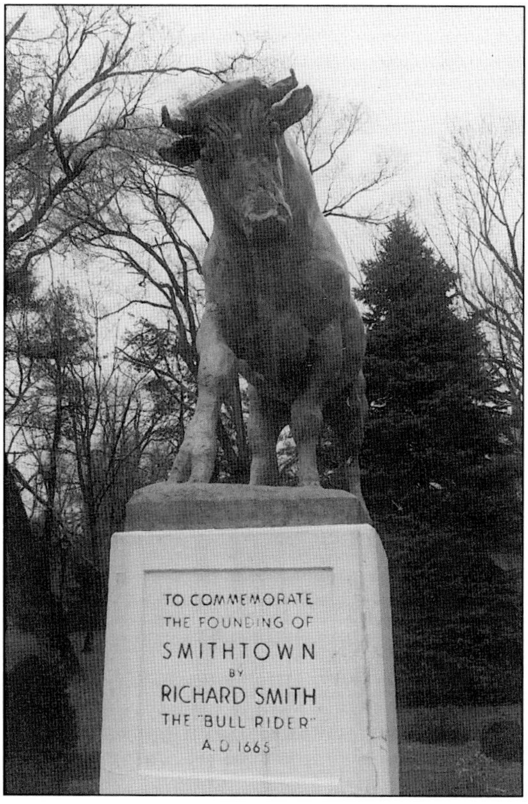

In 1903, Lawrence Smith Butler, a descendant of Richard Smith, proposed a memorable bull statue to his classmate, Ramsey. In 1923, the huge casting was complete but it would be two decades before Whisper, the bull, would arrive in Smithtown. In 1941, the town built the concrete pedestal for Whisper. Necessary funds were raised and the bull, weighing 5 tons, arrived via railroad and truck. Heirs of Ramsey donated it to the town of Smithtown on May 10, 1941.

Five

Picturesque Remnants of Nassau

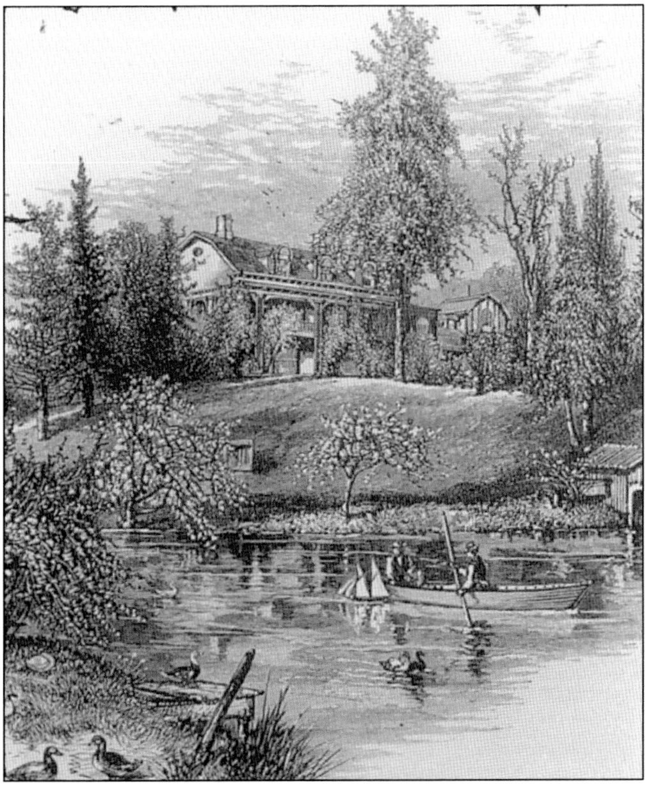

Cedarmere is located on a North Shore cliff overlooking Hempstead Harbor in Roslyn. William Cullen Bryant, a prominent 19th-century American poet, civic leader, and editor of *The Post*, bought the 40-acre estate in 1844. Known as Harbor Hill, it is the second highest elevation on Long Island. William Kirk, a Quaker farmer, constructed the oldest part of the house in 1787.

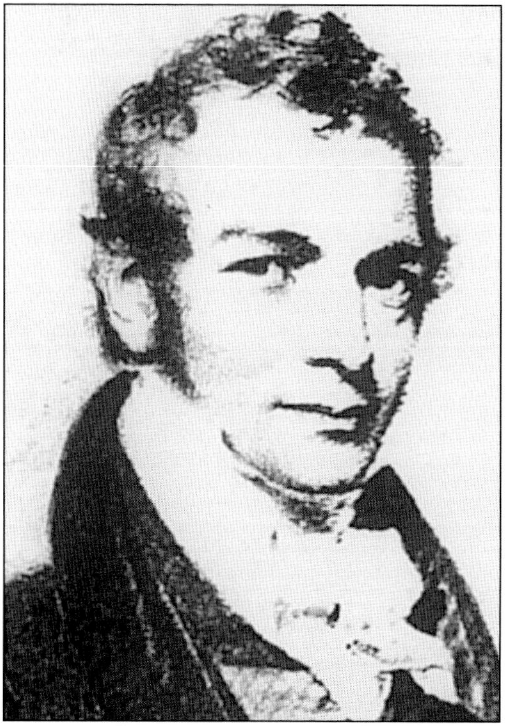

William Cullen Bryant was born in Massachusetts. He came to Long Island seeking solitude. He wanted a retreat where he could work on his poetry and enjoy his love of nature. *Thanatopsis* was written here. He transformed his estate into a horticultural manor when he planted many exotic trees and flowers. Bryant prevailed upon the officials of this quaint village to change the name of the hamlet to Roslyn.

Bryant's Library was a gift to community of Roslyn. He wanted people to have a meeting place for intellectual and social pursuits. He bought land south of his estate and erected "The Hall" for that purpose. He watched the centennial celebrations from this vantage point in 1876. The original Bryant Library was located on Bryant Avenue. In 1951, it was in the path of the new Northern Boulevard bypass. Bryant's Library was relocated to the War Memorial Building.

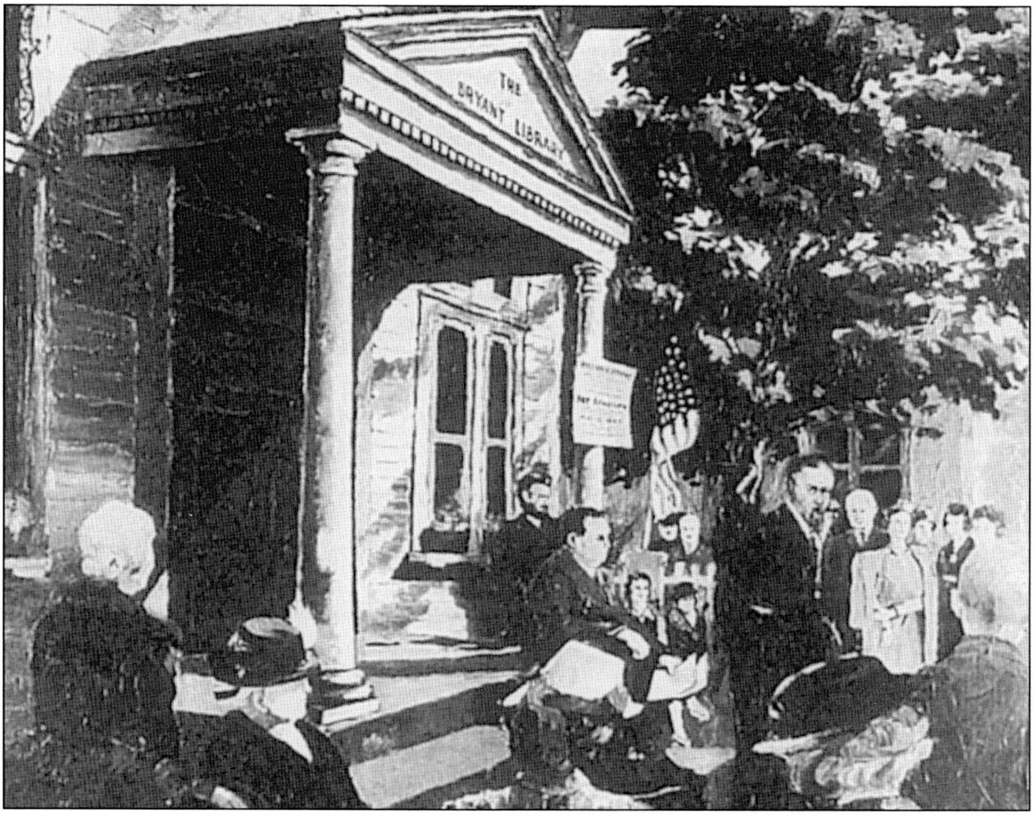

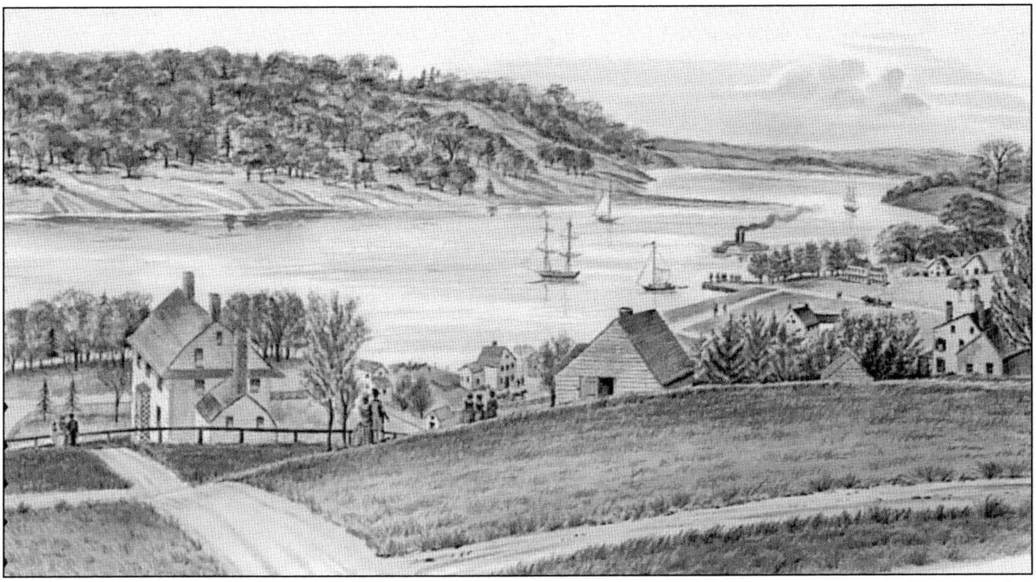

The deep water of Hempstead Harbor is free of oceanic waves. Bryant often sailed the 18 miles to his New York office. He founded the *Saturday Evening Post* and was its editor for 50 years. He also founded an American School of Journalism, creating a staff of correspondents to cover all cities and restoring public respect for the press. He suggested an uptown park be established in New York City. The result was Central Park.

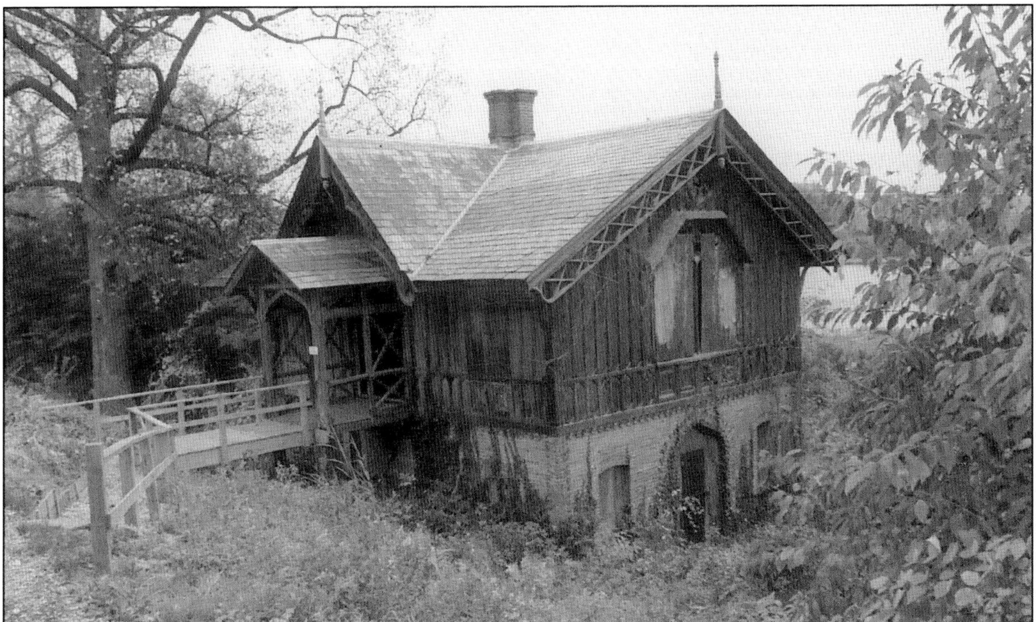

Bryant's Mill at Cedarmere was part of his estate. Located near the pond, it served the needs of the family. In 1901, the house and mill were destroyed by fire. The mill was rebuilt in Gothic Revival architecture and elaborately decorated on a brick foundation. It is the only Gothic Revival Mill on Long Island. The 7-acre estate was left to Bryant's daughter and grandson, Harold Godwin, who added the sunken gardens and bridge.

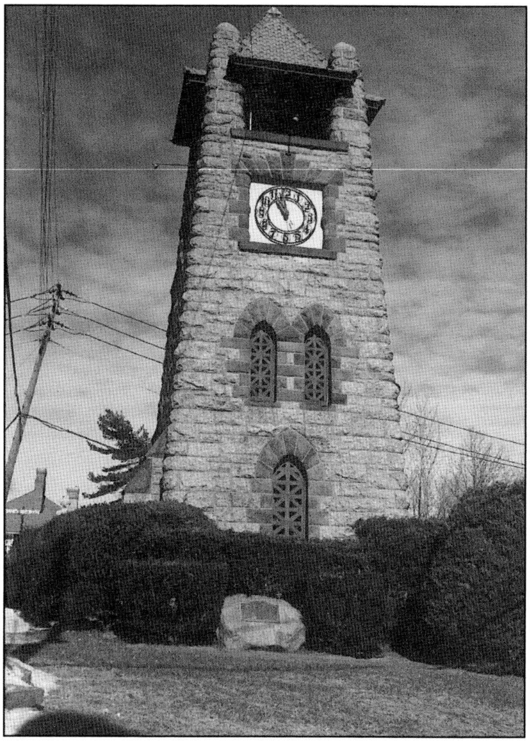

The children of Ellen Ward gave Roslyn Clock Tower to the community. Built in 1895 it faces Washington Manor Inn. President Washington sojourned there on his tour of Long Island. The clock tower is constructed from granite with red sandstone trim and rises 44 feet above street level. It is of Egyptian Architecture. Seth Thomas made the clock with metal dials 3 inches thick.

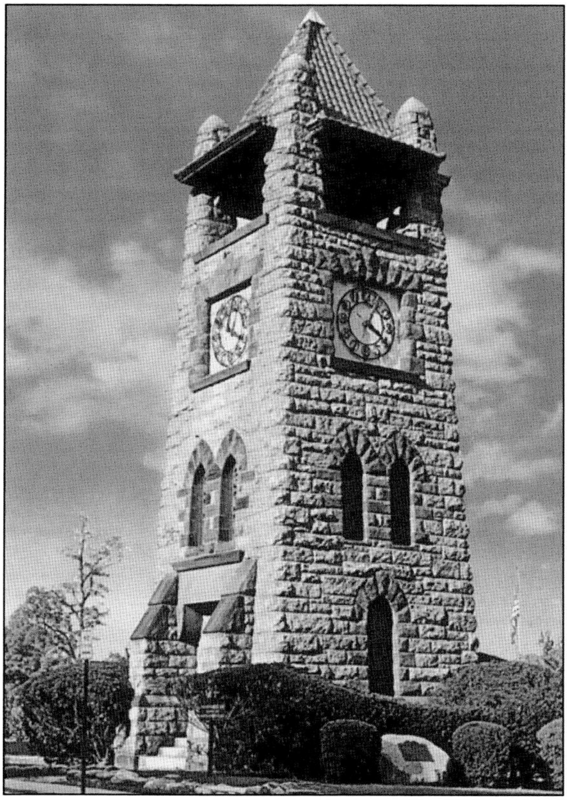

The bell in the Roslyn Clock tower was installed in November 1895. It weighs 2,500 pounds and strikes the hour. It was also used as the first fire alarm. The clock must be wound once a week. Caretakers find it easier to wind it halfway twice a week. It operates on large weights, which gradually descend from the mechanism to the base of the tower. The tablets surrounding the tower are dedicated to the servicemen of Roslyn.

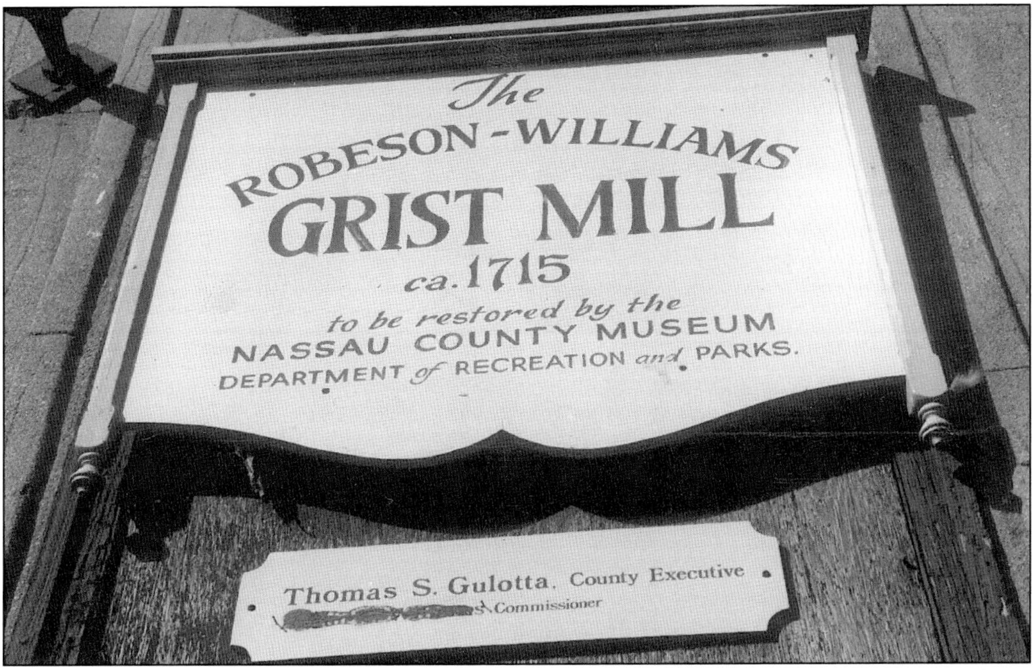

The Robeson-Williams Grist Mill in Roslyn was built by John Robeson on Old Northern Boulevard in 1706. He incorporated many Dutch framing techniques. In 1715, it was sold to Jeremiah Williams who enlarged it. Much of the 18th-century milling equipment and the 17th-century millpond have survived. The mill evolved into a charming tea house. It was acquired by the Nassau County Museums in the mid-1970s.

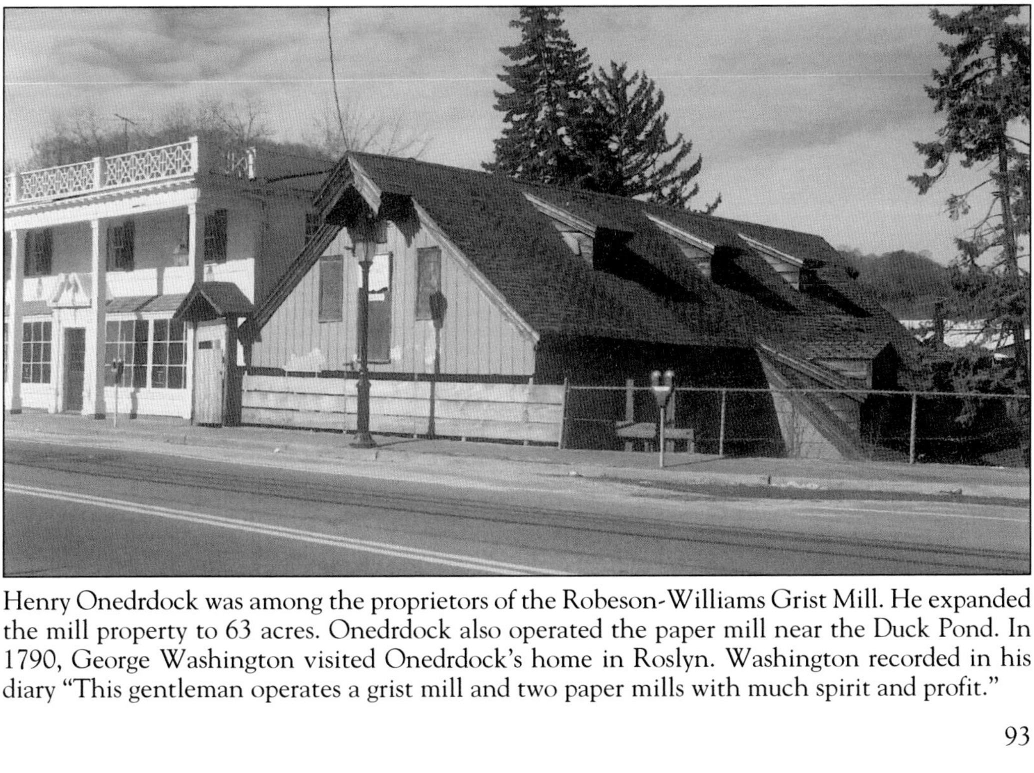

Henry Onedrdock was among the proprietors of the Robeson-Williams Grist Mill. He expanded the mill property to 63 acres. Onedrdock also operated the paper mill near the Duck Pond. In 1790, George Washington visited Onedrdock's home in Roslyn. Washington recorded in his diary "This gentleman operates a grist mill and two paper mills with much spirit and profit."

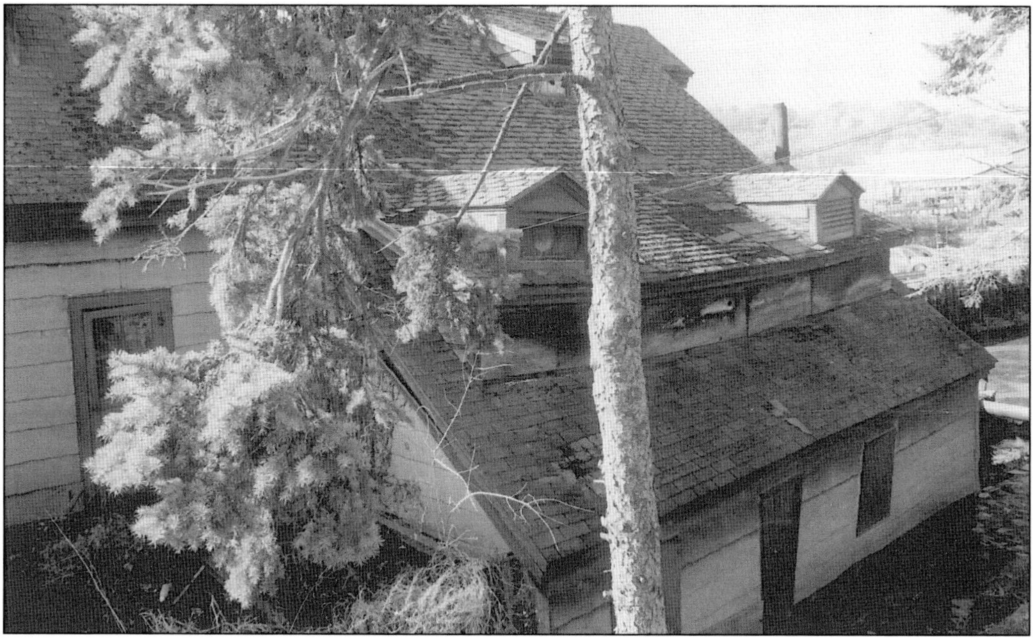

The Roslyn Board of Trustees restored the Robeson-Williams Grist Mill as a museum. Alice Titus, the clerk, displayed relics of the past such as Quaker bonnets, lace parasols, a British Red Coat, and hand-knitted items of two centuries ago. She served tea as an attraction. This evolved into the quaint and charming Roslyn Tea House. Many people dined there, including Roslyn's author and journalist, Christopher Morley. Garden luncheons were the order of the day.

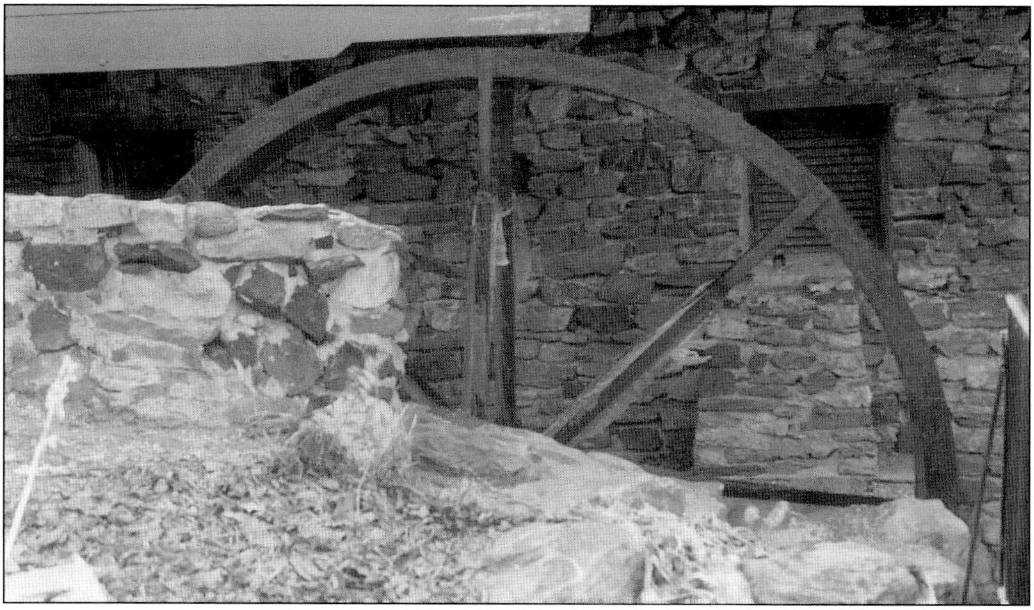

A replica of Roslyn Paper Mill stands where the original was built over two centuries ago. It is located in a park near the gristmill. In 1790, Henry Onderdonk invited President Washington to breakfast. He entertained him by showing him the operation of the paper mill and presenting him with a sheet of paper made in the mill.

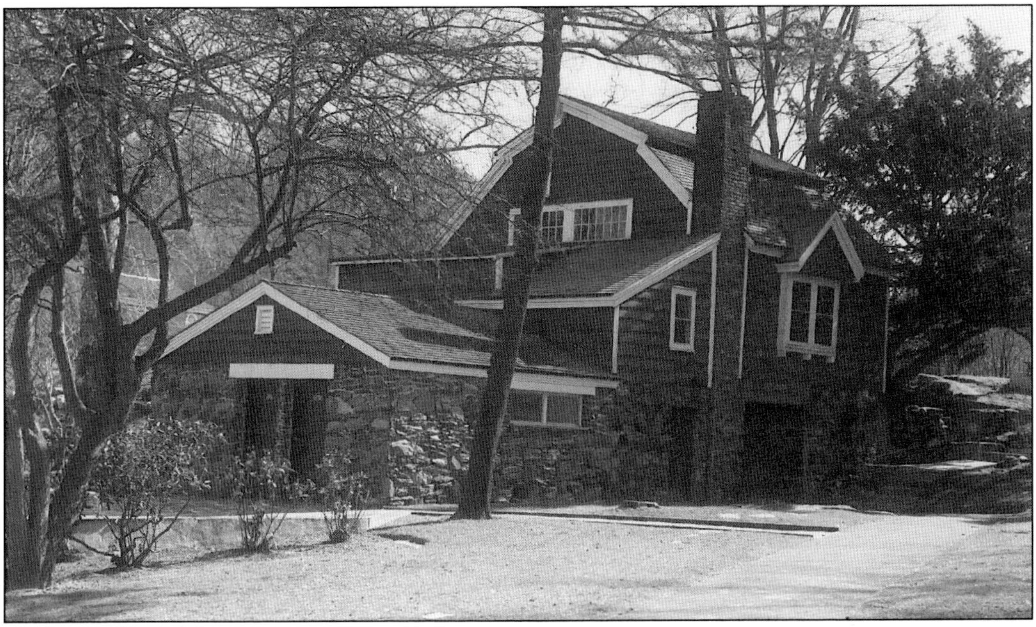

Harold Goodwin restored Roslyn Paper Mill in 1744. He established a park surrounding the mill. Paper Mill Road leads to the historic building. The northeast side bears the mill wheel and millrace where the water pours over the wheel. A tablet was placed in the mill attesting to Harold Goodwin's generous gift and his love for Roslyn. It was the first paper mill in the state of New York and was visited by President George Washington in 1790.

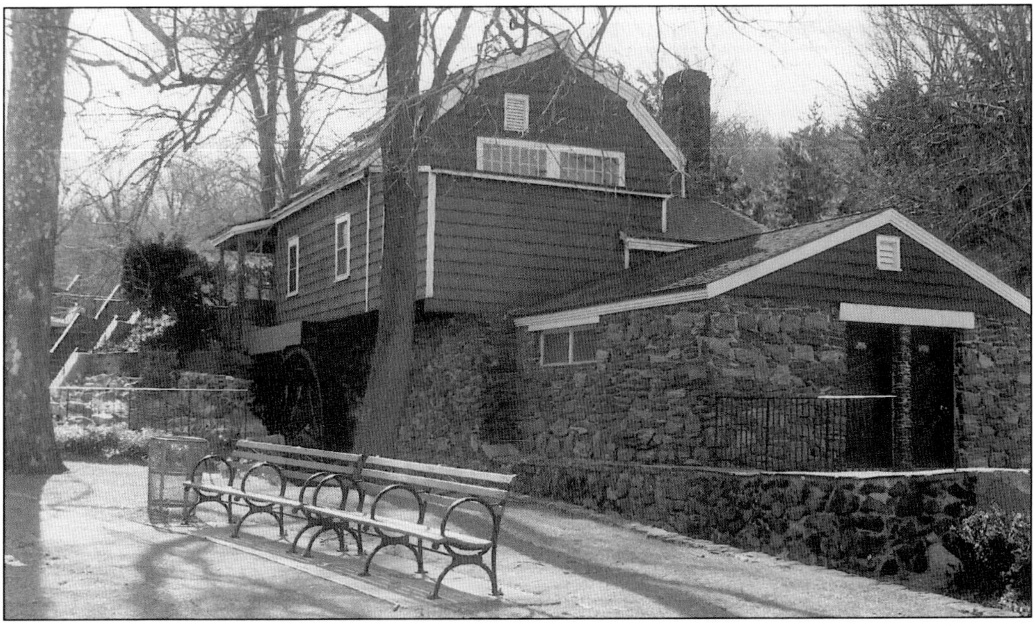

Roslyn Paper Mill operated throughout the American Revolution. The generating plant was on the lower floor and a meeting room on the upper floor. Paper making was a very profitable business. It was initially made from old rags. In 1801, Caleb Valentine bought the mill and it remained in the Valentine family for 90 years. In 1880, steam power was added. The mill also manufactured straw board, made from straw.

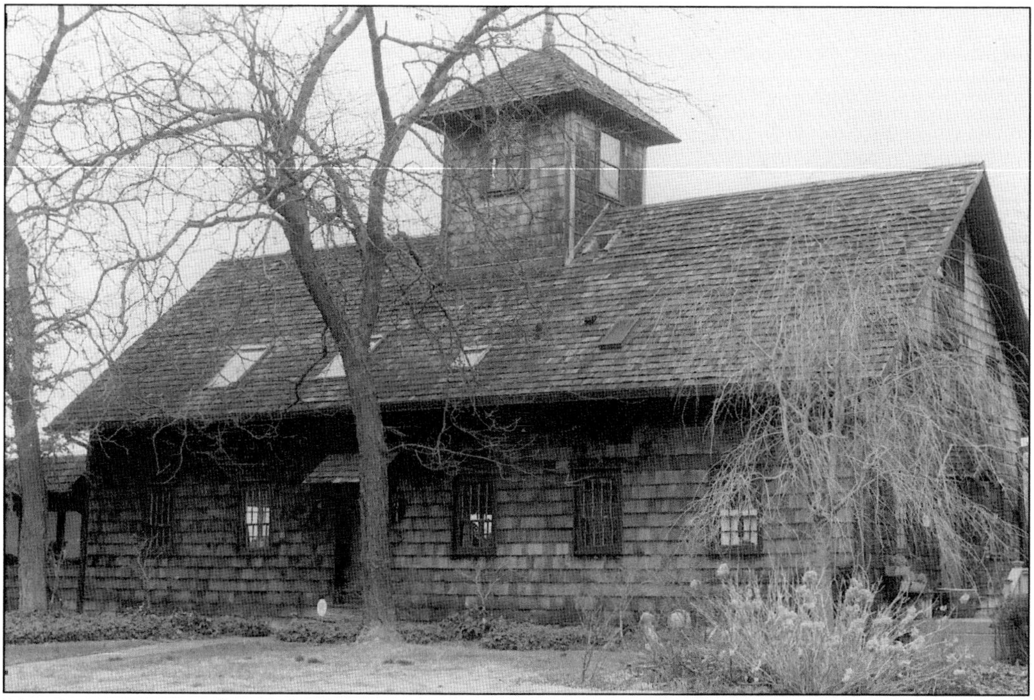

Plandome Grist Mill was built in 1693 on Manhasset Bay. It was operated by the tide that rushed through Leed's Pond. Last used to grind grain in 1906, its large waterwheel was lost in a storm. In 1951, it was moved back 70 feet and converted to a residence. Oak beams support the frame and cupola. The millstones and wooden gears are still in place.

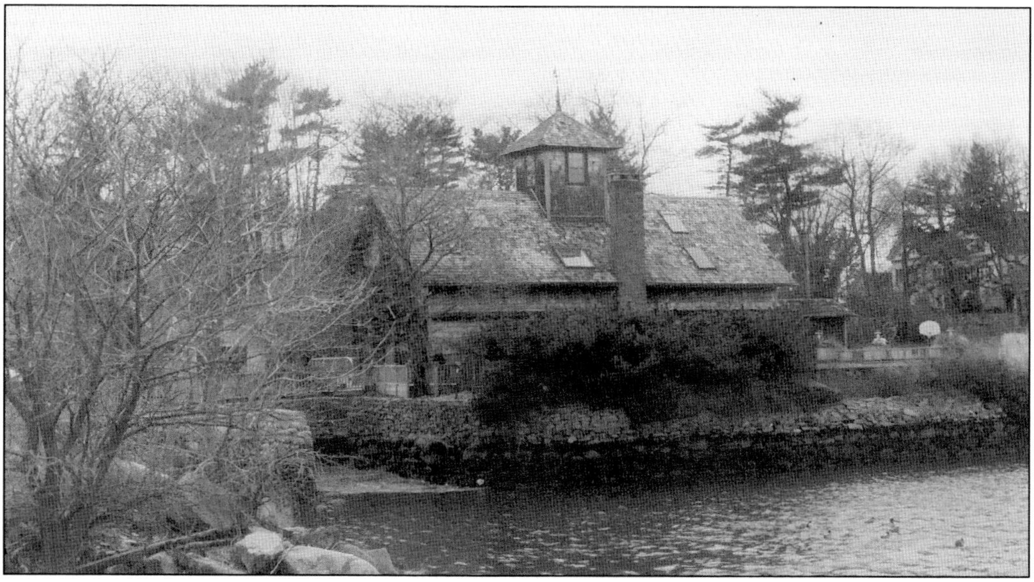

The Plandome Mill has been a residence since the mid-20th century. A picture window was installed to enjoy the scenery. The wooden grain hoppers and French limestone wheels that ground rye, corn, and oats are in the building. Though there have been no structural changes, a partial wrap around deck has been built on the outside. Devotees of antiquity are happy to see so much of the original mill intact on Manhasset Bay.

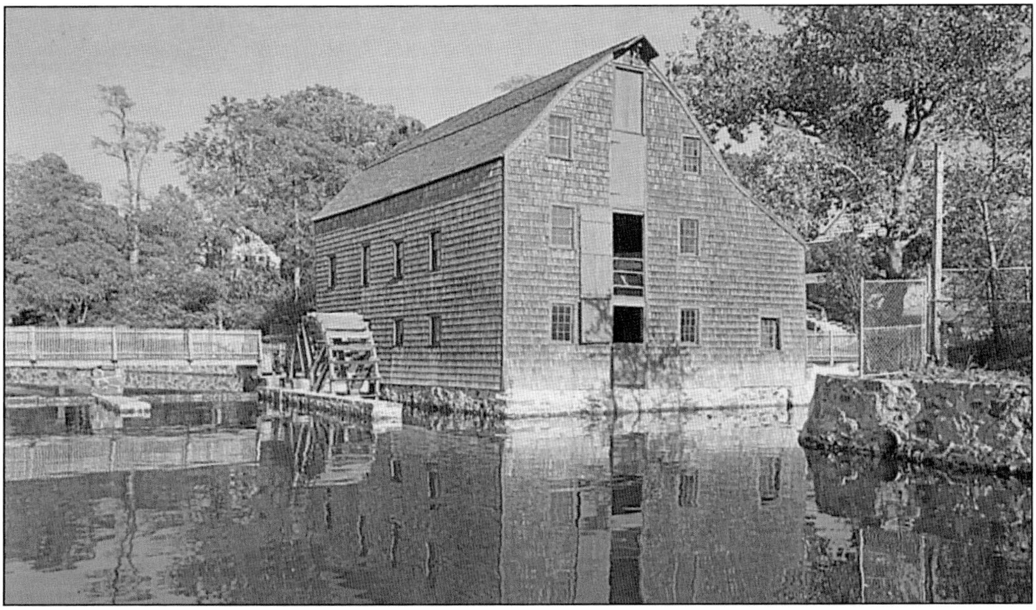

Saddle Rock Grist Mill is a landmark. It is a simple, picturesque mill built before 1700. Power for the mill is derived from the ebb and flow of the tide. Great Neck was once called "Madnan's Neck." It seems there was an aggressive woman, Nan Heatherton, who landed with some followers at Kings Point and laid claim to the entire area. She fought off all contenders with a belligerence that earned her the title "Mad Nan"—hence, Madnan's Neck.

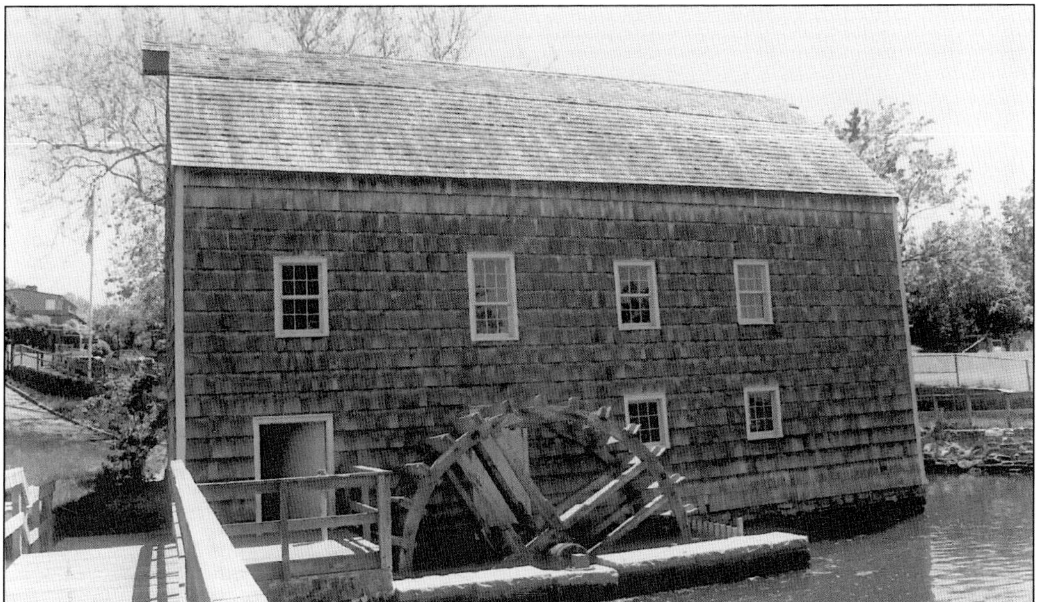

Milling was operational on the Great Neck Peninsula in 1679. Robert Hubbs sold the mill to Henry Allen in 1702. Allen was the first to begin keeping records. He worked the mill for many years. His son, David, left his flour barrel brand mark on the mill door. Milling was revolutionized in the early 1800s when Oliver Evans invented automotive devices that relieved the miller.

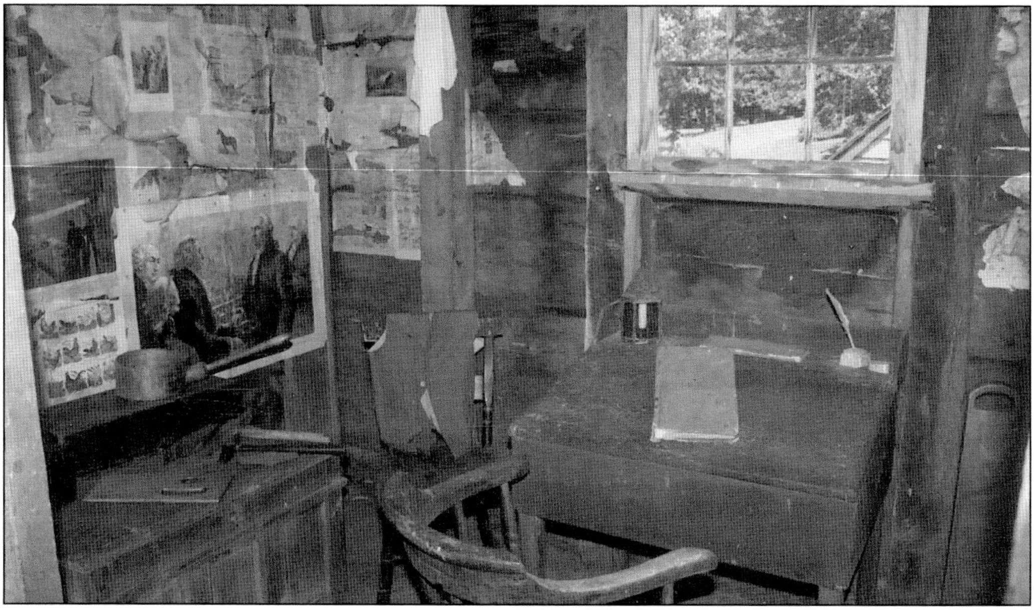

The interior of Saddle Rock Mill has a rustic office. Oliver Evans improved mill technology when he installed elevators and conveyor belts to move the grain. His ideas were introduced into many mills around the country. Rumbling gears, whirring shafts and moving belts eased the miller's work. Oats, Indian meal, corn, chicken feed, bran, buckwheat, rye, and flour were often milled and shipped by steamboat to New York City markets.

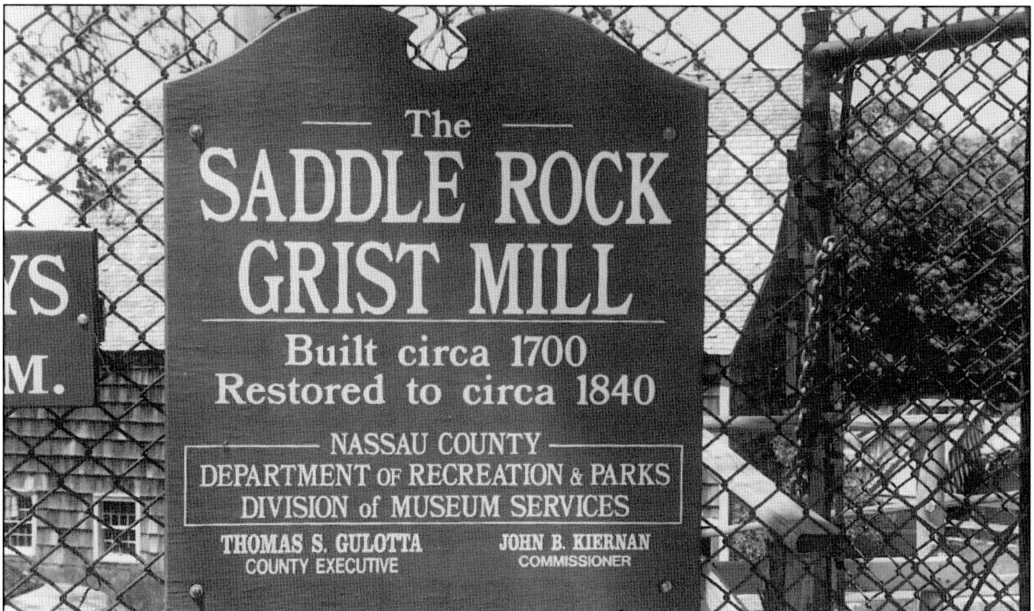

Saddle Rock Mill became a terminal for the shipping business. In 1829, the Udall family bought the mill and operated it until 1870 when grain centers opened and new roller mills rendered the stone mills obsolete. A little village had sprung up around the mill, which had been the center of commerce. Sloops docked beside the building and carried farm produce to the city, returning with manufactured products.

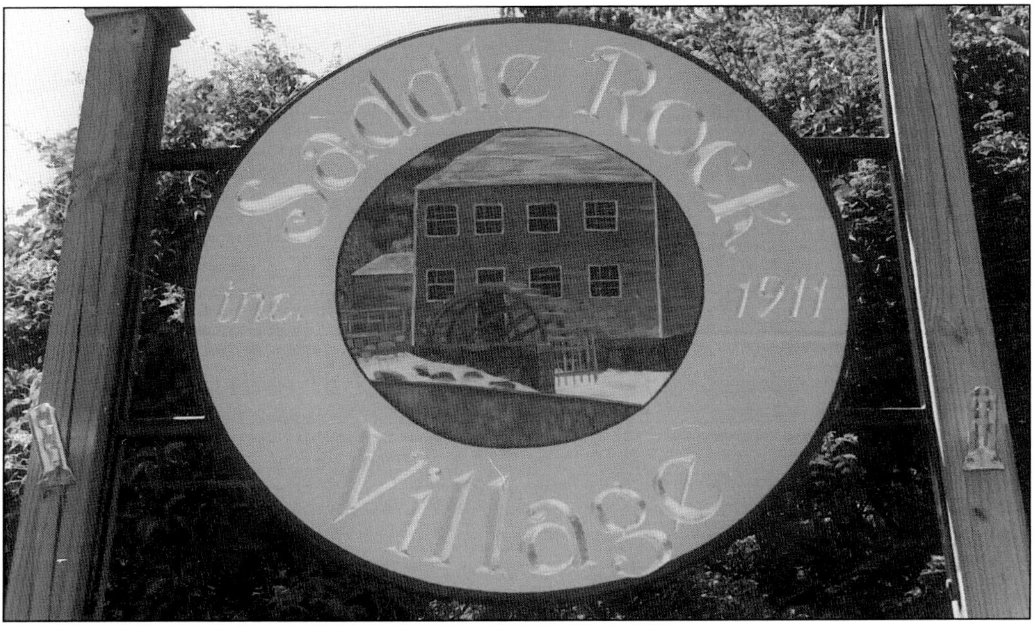

Louise Udall Eldridge owned Saddle Rock Grist Mill. She had the mill restored and installed electric motors in 1940. The building is a working example of the chief enterprise of rural Long Island. In 1955, the mill became the property of Nassau County as an historic site. In 1870, a barrel of flour was $12, a bushel of corn was $1, and Indian meal cost 2¢ per pound.

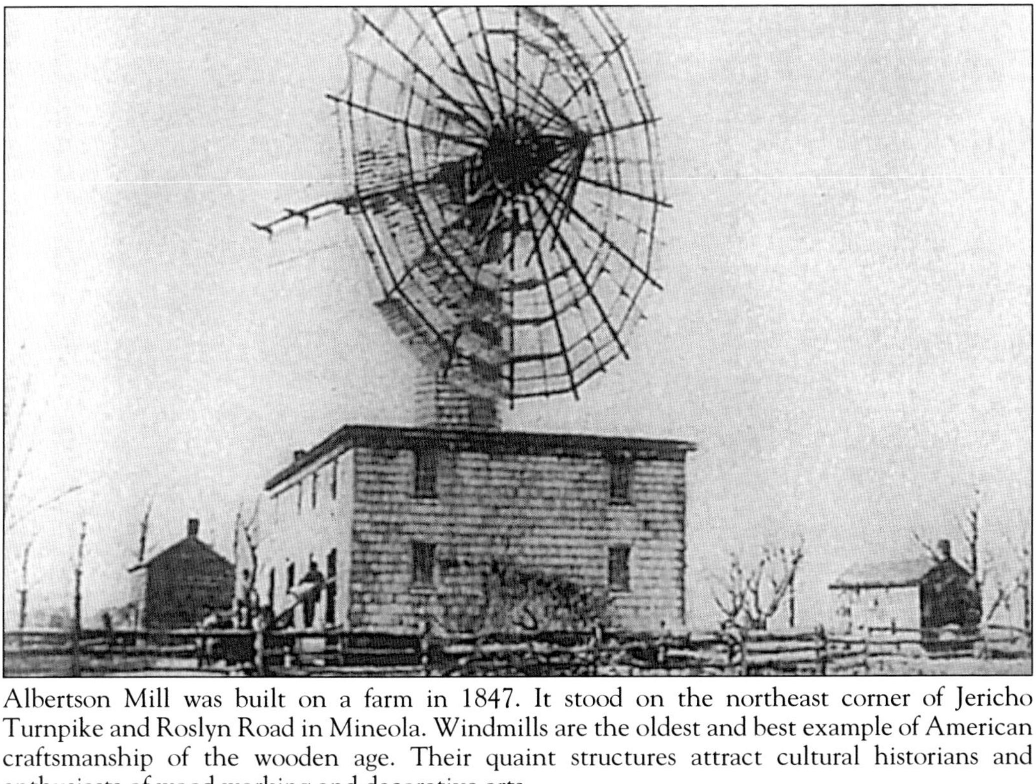

Albertson Mill was built on a farm in 1847. It stood on the northeast corner of Jericho Turnpike and Roslyn Road in Mineola. Windmills are the oldest and best example of American craftsmanship of the wooden age. Their quaint structures attract cultural historians and enthusiasts of wood working and decorative arts.

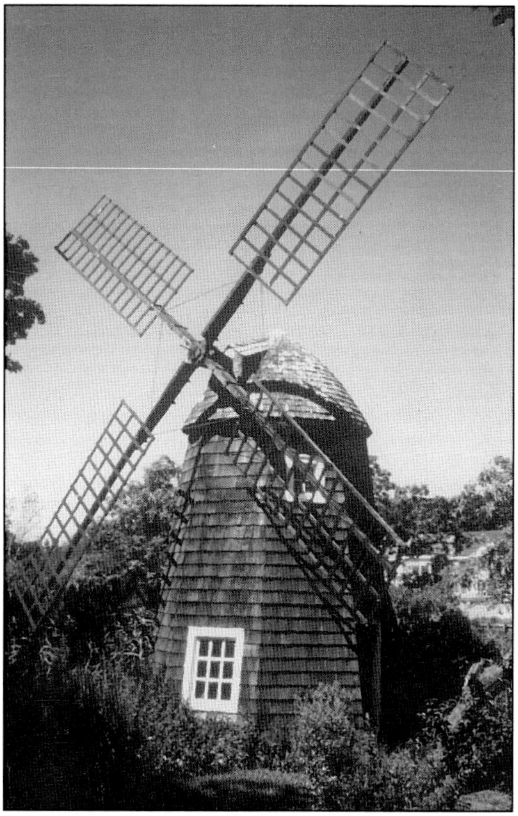

Hewlett Mill was built in 1791 on the south shore of Long Island. It operated on the west side of Seawave Drive as a flour mill for many years. The Hewlett family settled on the Rockaway Peninsula in 1636. They were among the first pioneers to establish themselves along this sandy stretch of Rockaway that juts into the sea.

Hewlett Hotel stood on the corner of Main and Front Streets since 1797. It was a celebrated inn for over two centuries. The Hewletts acquired and bought other properties including Rock Hall in Lawrence. Josiah Martin, a planter from the West Indies, built this Georgian estate in 1767. The Hewlett family bought it in 1814 and maintained it for many years.

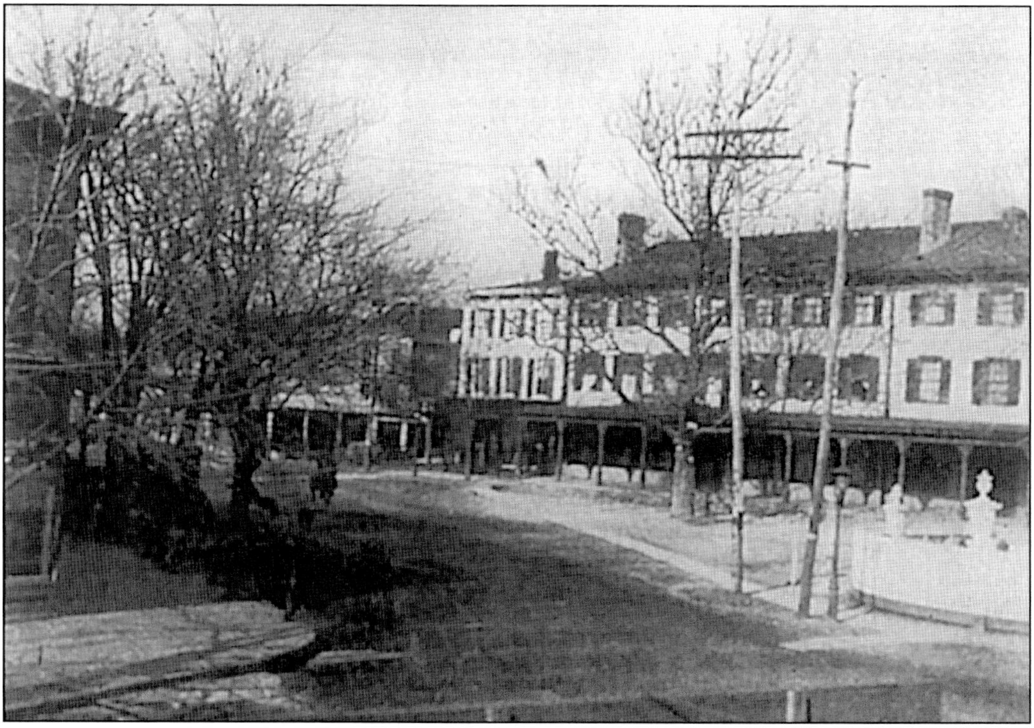

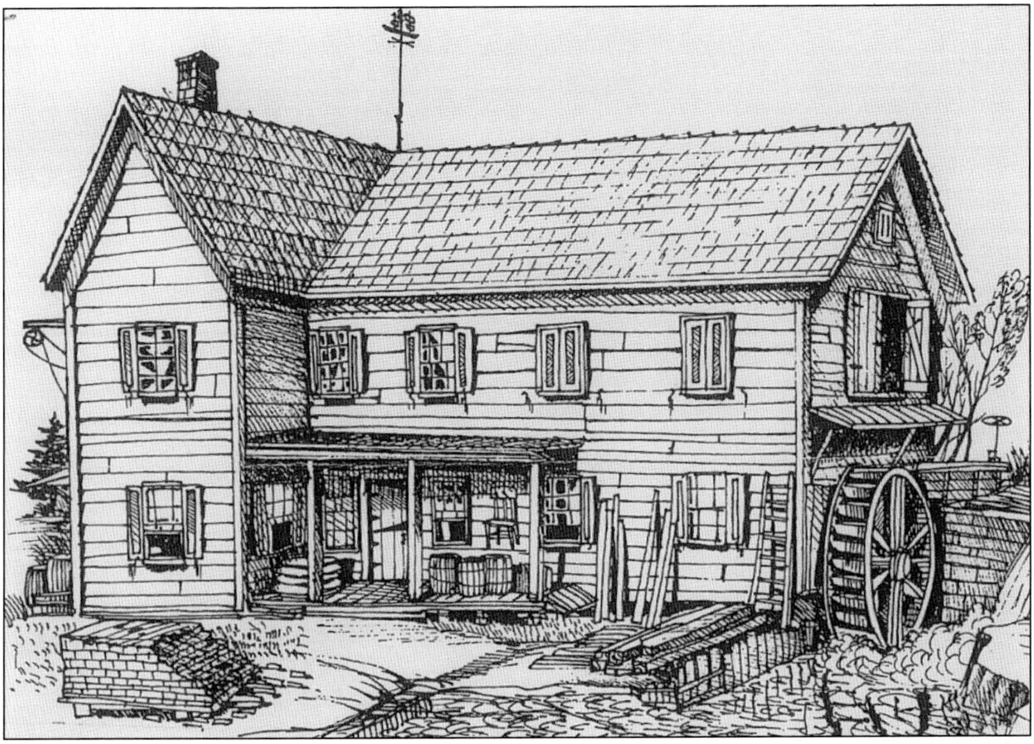

Joseph Haviland erected the gristmill in East Rockaway in 1688. It was a tidewater mill and the hub of business and social activities as it was the only center of shipping and milling in Rockaway. In 1818, Alexander Davidson replaced the great wheel with an underwater turbine. Rockaway Mill has become an extensive museum and is now in the National Register of Historic Places.

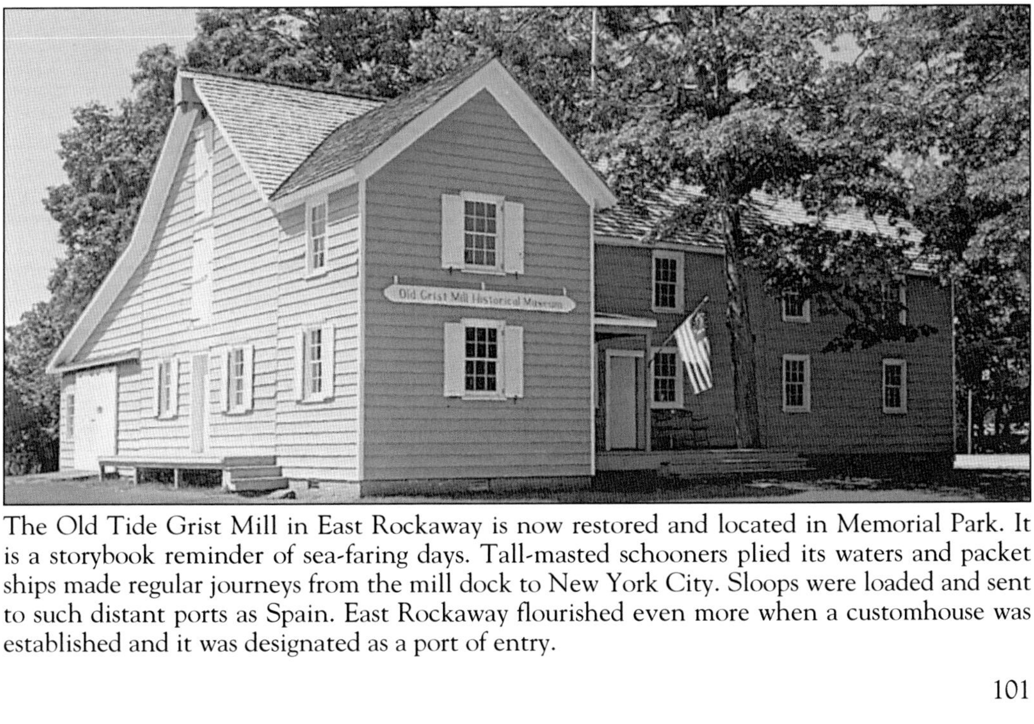

The Old Tide Grist Mill in East Rockaway is now restored and located in Memorial Park. It is a storybook reminder of sea-faring days. Tall-masted schooners plied its waters and packet ships made regular journeys from the mill dock to New York City. Sloops were loaded and sent to such distant ports as Spain. East Rockaway flourished even more when a customhouse was established and it was designated as a port of entry.

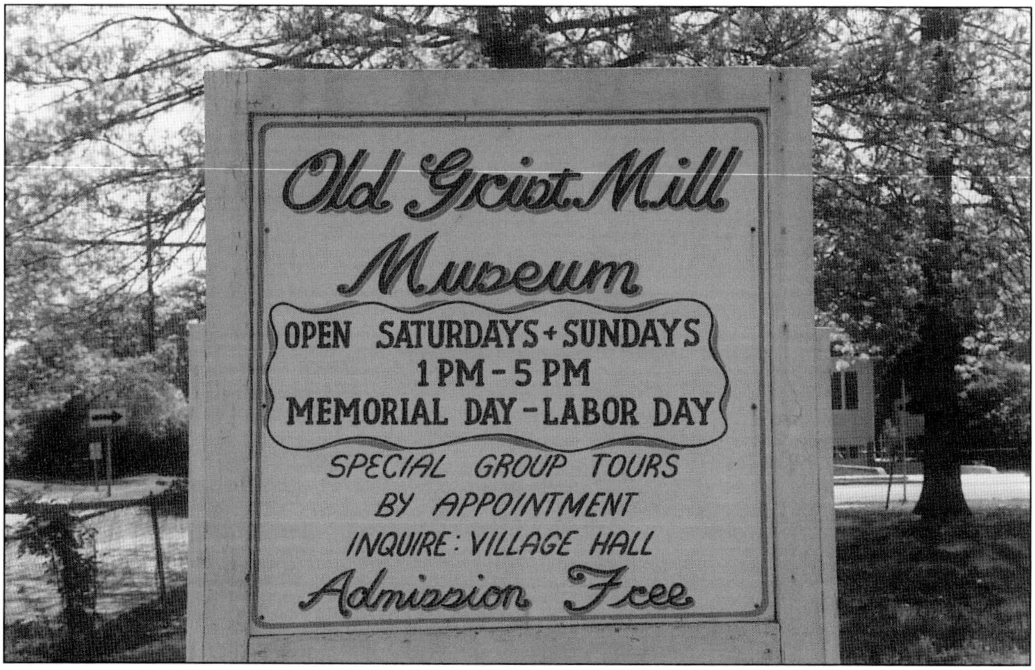

Alexander Davidson bought the Old Tide Grist Mill. He straightened the river to run parallel with Ocean Avenue. He then moved the mill north and had Atlantic Avenue lain out. In later years, the mill became a sawmill and The Davidson Lumber business was established here. The gristmills of yesteryear had their own trademarks. A crow's foot was the trademark of 17th-century gristmills.

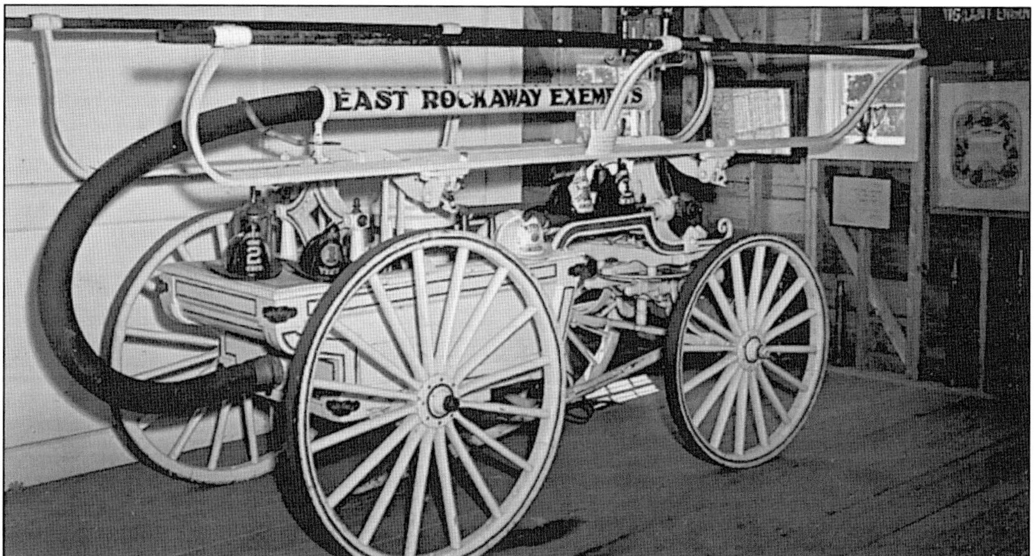

The first hand pumper in East Rockaway was named "Tootsie." It was pulled and operated by courageous volunteer firemen. The pumper is now preserved at the gristmill as part of the historical collection. The museum today is a rare example of a late 17th-century mill that retains the characteristic framing technology of 1688. The mill has been moved several times; however, the building retains architectural associations with East Rockaway of another era.

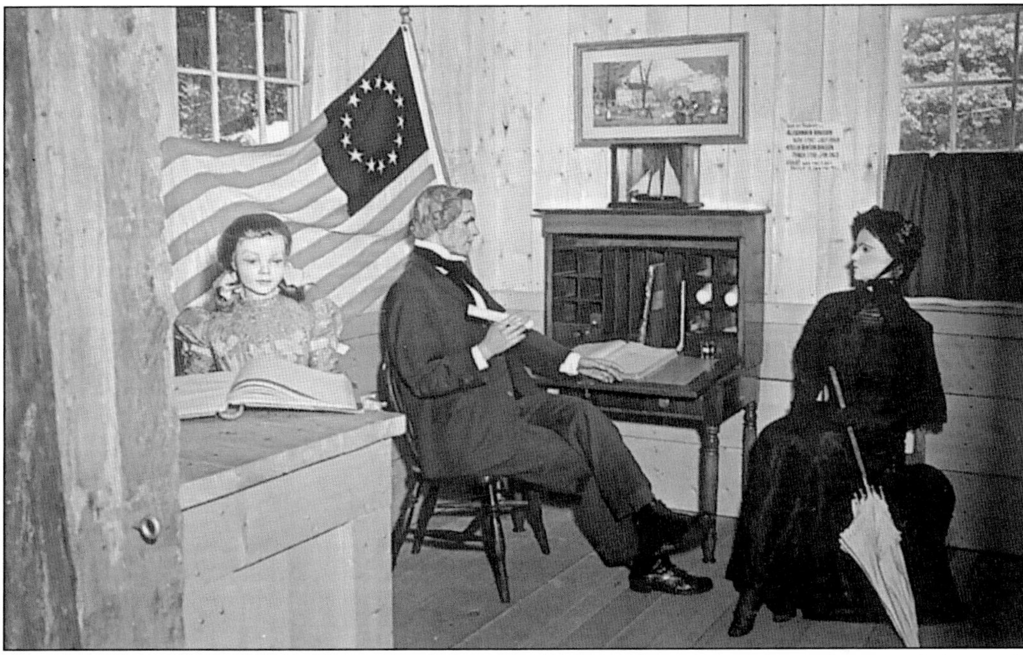

Alexander Davidson, son of the original owner, discusses the state of the country with his wife, Amelia, in early 1800. The restored mill office continued to serve in that capacity long after the mill ceased operating. The tidewater mill was located on Long Lane, now Ocean Avenue. It prospered and passed through several owners before Alexander Davidson bought it in 1818. In 1962, when the A&P Company bought land, East Rockaway acquired the mill and the historical society was formed.

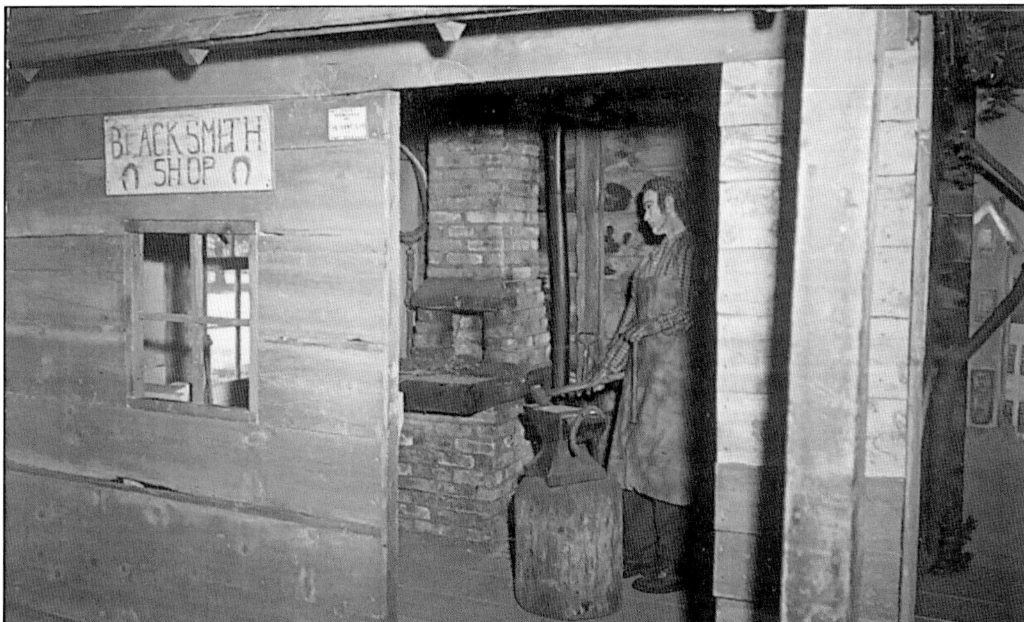

The blacksmith shop was vital to early settlers. In East Rockaway it was opened at the mill. It was here that horses were shod, wagons repaired, boat parts fixed, and household items made.

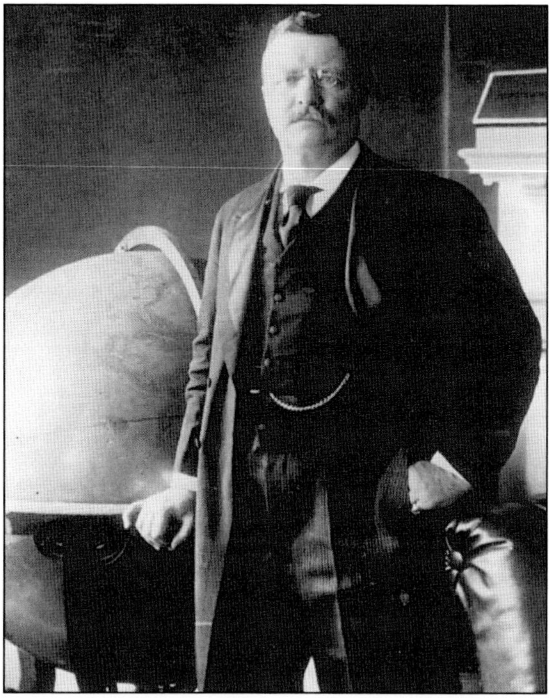

Theodore Roosevelt was a vice-president candidate on the 1900 Republican ticket with President William McKinley. Roosevelt became the 26th president when McKinley was assassinated in 1901. The acquisition of the Panama Canal Zone was his greatest achievement. His concern for two little bear cubs brought about the beloved Teddy Bear. While dining at the Maxwell House Restaurant in Nashville, Tennessee, he was offered a second cup of coffee. He accepted with the quotation, "It's good to the last drop."

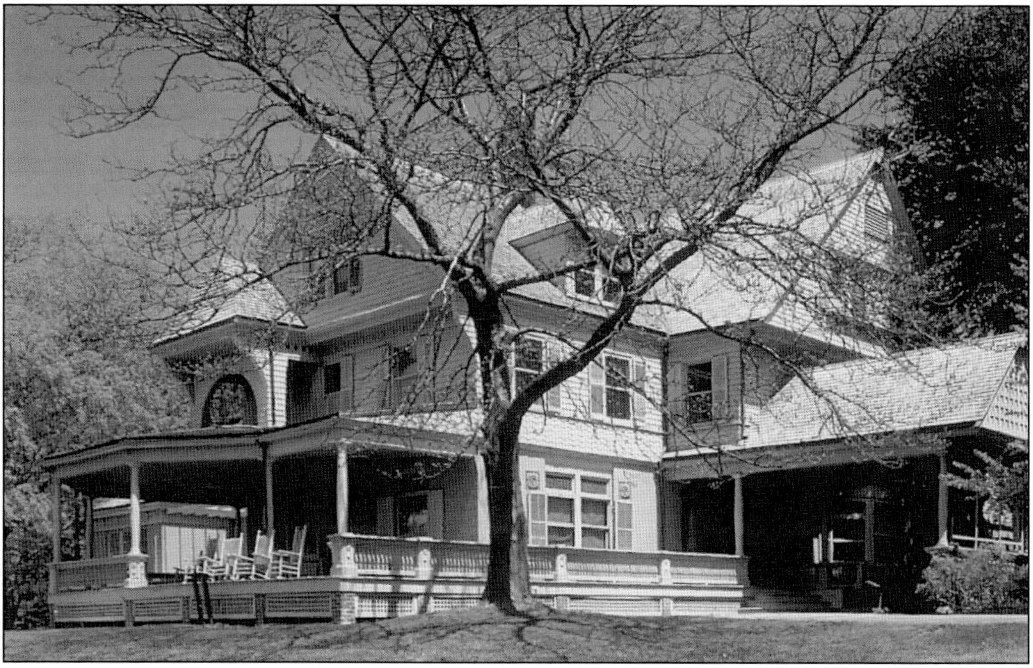

Sagamore Hill was established by the president's father, as a summer residence, in 1872. Theodore spent vacations exploring the fields and woodlands around Cove Neck. A decade later he bought the property and built a huge 23-room Victorian mansion in 1884, while he was a member of the New York State Assembly. The solidly built rambling mansion contains many original furnishings and artifacts of Roosevelt's hunting days. This was the President's summer Whitehouse from 1901 until 1908.

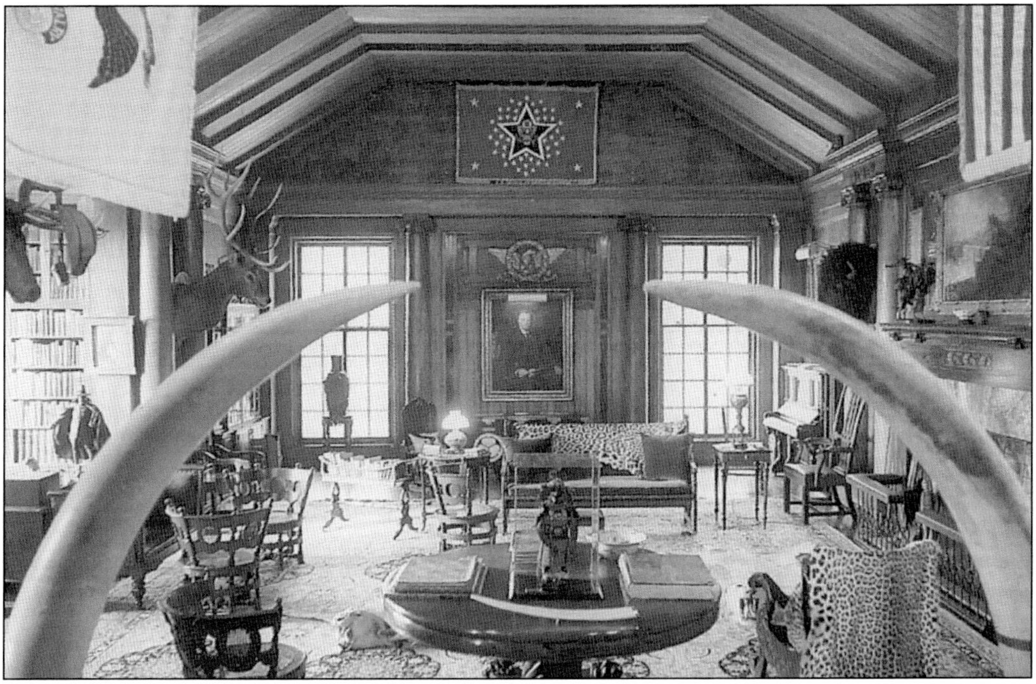

The North Room at Sagamore Hill is filled with trophies, books, paintings, flags, and guns. Roosevelt's office was the drawing room of his mansion with its big polar bear rugs and deer heads. He was noted for his conservation of natural resources. Note the huge antlers in the foreground and deer heads on the wall. Most of the furnishings throughout the house are original pieces.

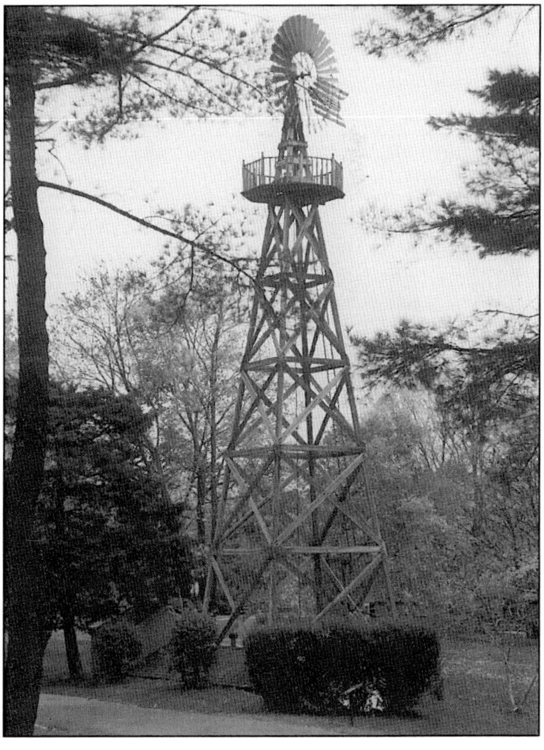

Sagamore Hill Windmill stood on the Roosevelt estate in the town of Oyster Bay. It is surrounded by rolling lawns and wooded terrain overlooking Hempstead Harbor. Oyster Bay is both a town and a hamlet. The town spans 104-square miles from the ocean to the sound. Sagamore Hill was visited by dignitaries from every occupation while Roosevelt served for nearly two terms as President of the United States.

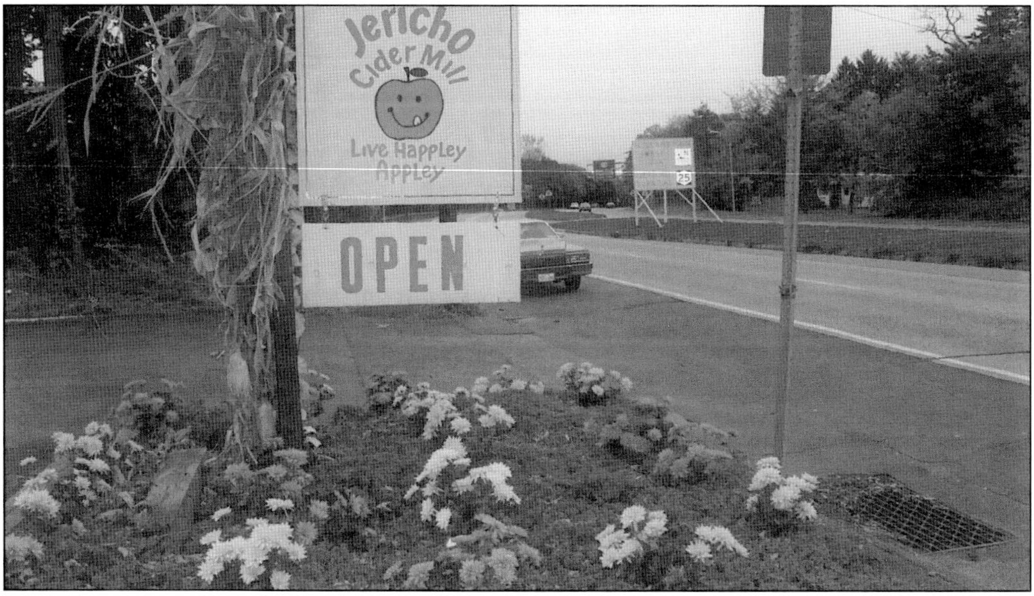

Jericho Cider Mill was built in 1855 on a hill in Jericho. Tunnels were cut through the hill to store aging cider. The mill produced champagne cider until Prohibition, when sweet cider became popular. The process utilized whole apples and there was a plentiful supply of cider for home use, trade, and commerce. The apples went through a washing, chopping, squeezing, noisy process in a clanking symphony of sound.

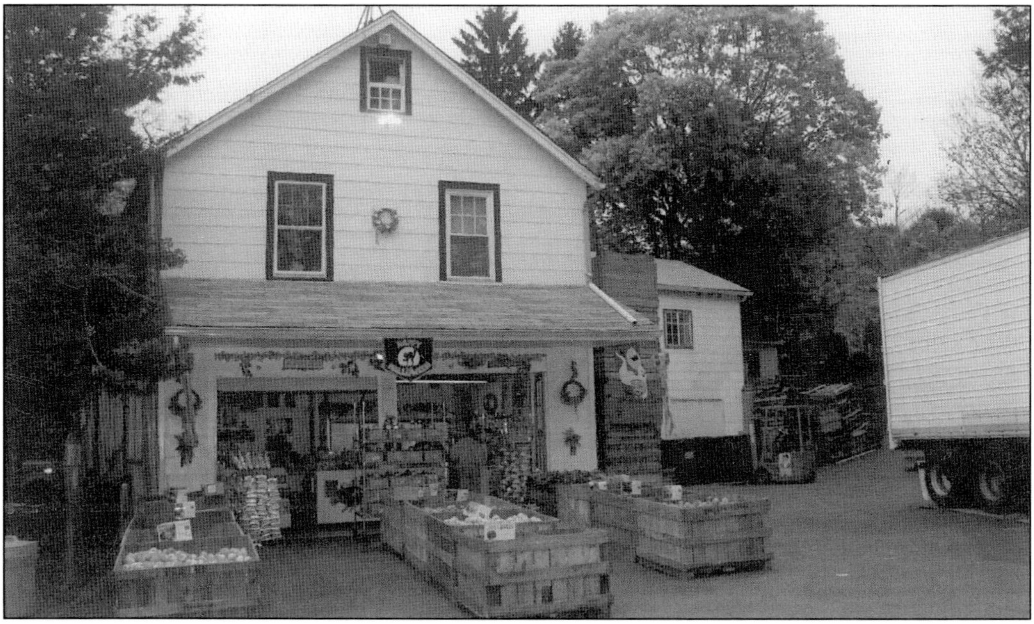

The Jericho Cider Mill converts apples into cider. This provided a preserved beverage that could be stored under proper conditions indefinitely. The cider mill was enveloped in ahearty aroma of mashed Macintoshes, Romans, Northern Spies, Cortlands, Red Delicious,and Russetts that were the pride of cider merchants. The mill is situated behind the white retail outlet. It is the oldest cider mill in the country, producing 100,000 gallons of sparkling cider per year.

Six

Nature's and Man's Ingenuity

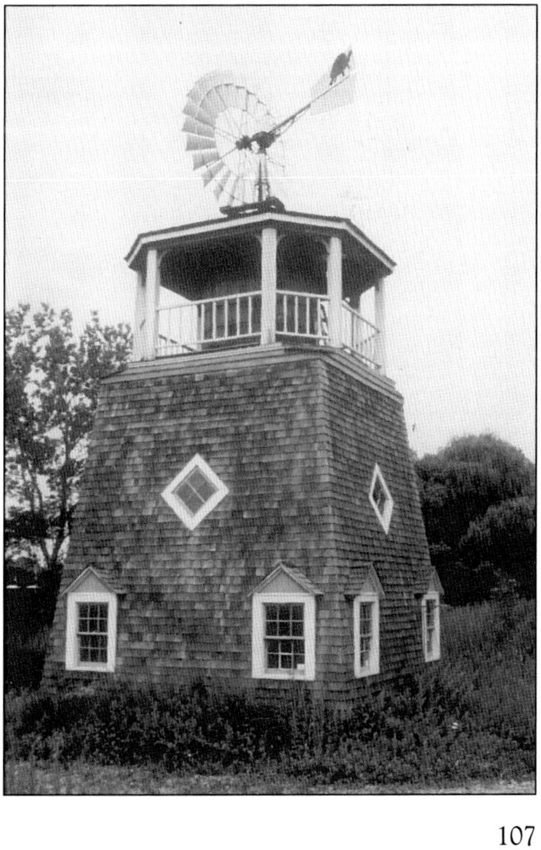

Douglaston Water Pump Mill was built on Arleigh Road in Douglaston. It is a remnant of the Van Wyck estate. It was used to pump water for land irrigation and to supply the manor house. The 225-year-old squat windmill has become a private home. In the mid-1850s, the George Douglas family purchased Van Wyck Manor. The mill front affords a view of a giant old oak tree that predates the mill.

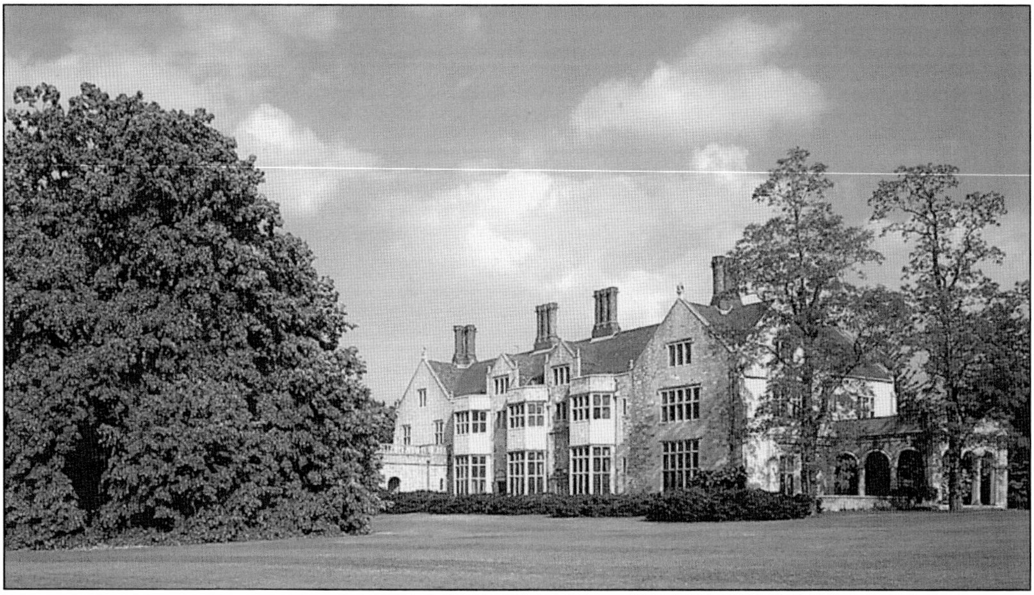

Coe Hall and Planting Fields evolved when William and Mai Coe purchased the estate in 1913. Their 65-room, Tudor Revival mansion was built between 1918 and 1921. The original house burned and the new building had to be carefully fitted into the site of the former house so as not to disturb the trees and shrubs. Coe was an insurance broker who married Mai Rogers, daughter of the H.H. Rogers of the Standard Oil Co.

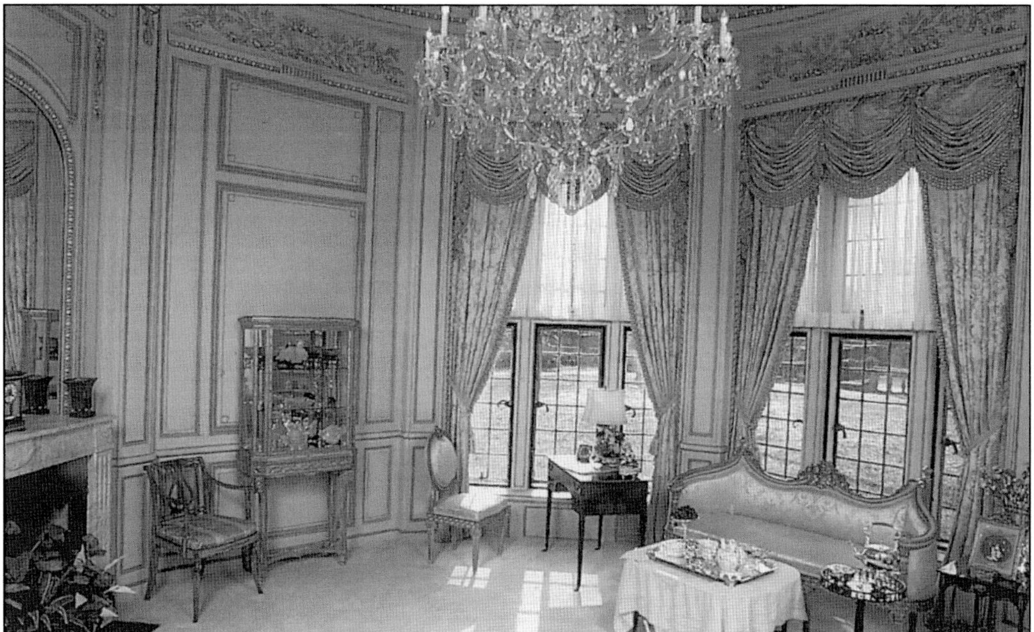

The reception room of the Coe mansion reflects the Gold Coast era. Situated in a Louis XVI setting, Mai Rogers Coe entertained guests in this room. A 500-acre estate comprised fields of extraordinary plantings and formal gardens. It was one of the finest botanical gardens in the country. A charming tea house is situated among the exotic plantings. Trees, shrubs, and flowers have been imported from around the world.

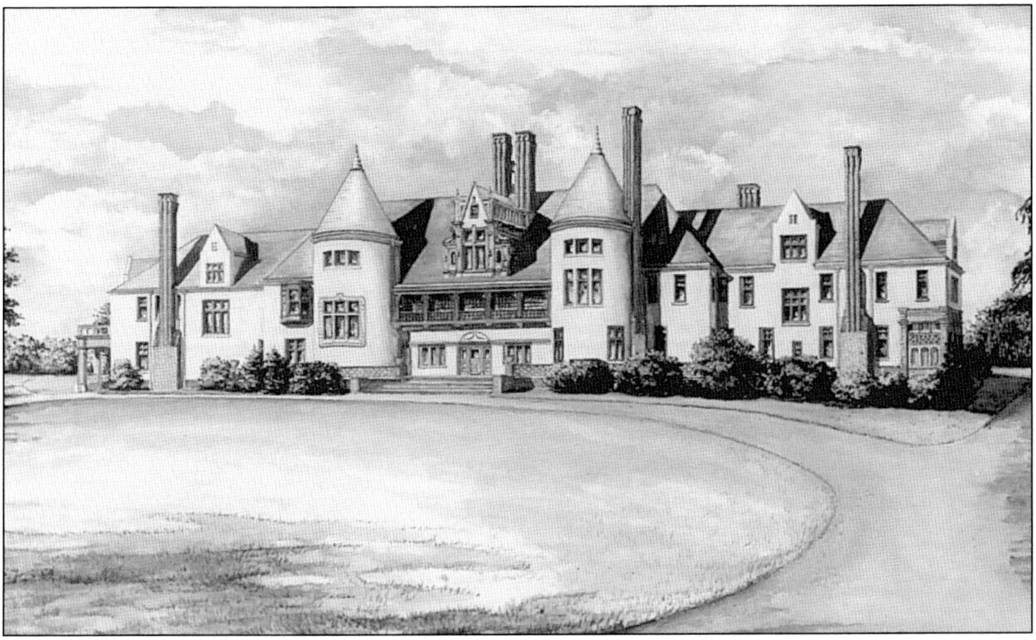

Coindre Hall Park is a 33-acre Gold Coast waterfront estate on a bluff overlooking Huntington Harbor. Built by George McKesson Brown in 1906, it was originally called West Neck Farm and was copied from two French castles. Brown was president of a pharmaceutical firm. During his constant trips to Europe, he collected furnishings for his palatial home. It is best viewed from Huntington Harbor. The family lived on the estate long after it was sold.

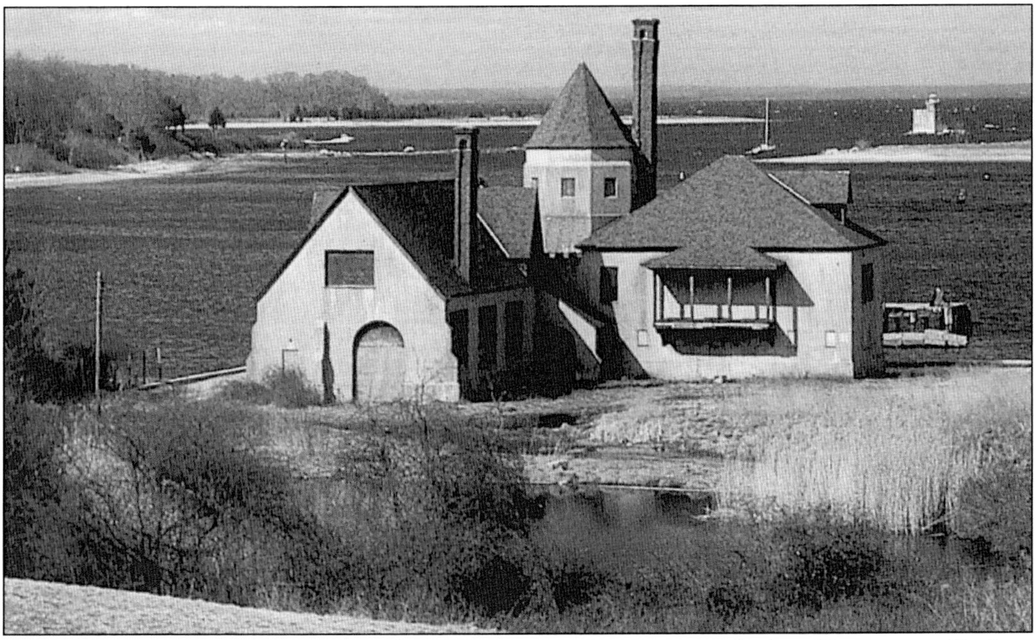

Coindre Boat House was constructed in 1912. Situated on a private dock in Huntington Harbor, it once housed an imposing yacht. The boathouse was built as a miniature replica of the main mansion. It was embellished with trophies and nautical equipment. Boathouses were a popular addition to Gold Coast estates on the waterfront.

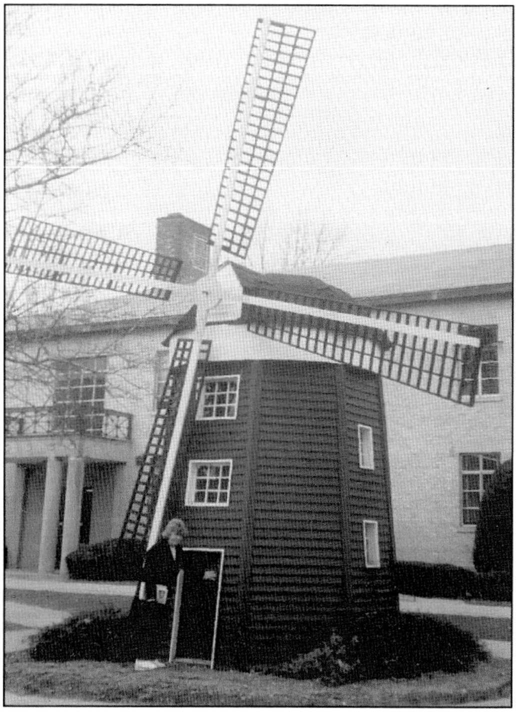

Hofstra Mill was part of the estate of William and Kate Hofstra. It was built in 1904 as a country estate on 13 acres. Hofstra was a millionaire in the lumber business. He was president of the Nassau Lumber Company. She furnished the house with fine old furniture and he landscaped the estate with trees from around the world. He died in 1932 and she a year later. They had stipulated that the estate be used for education or public good.

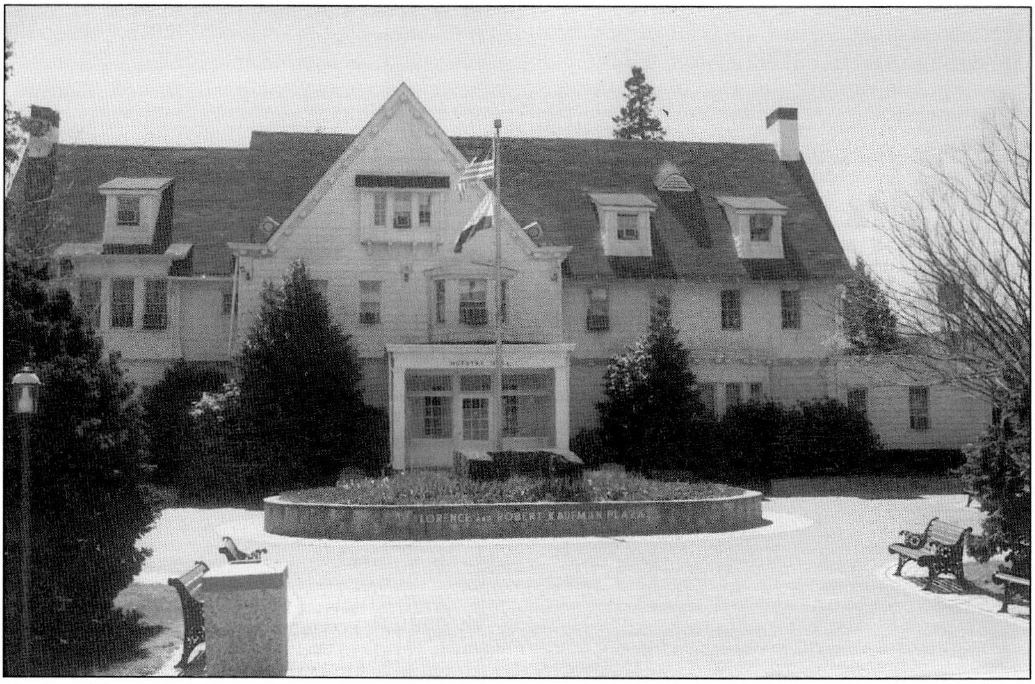

Hofstra University was conceived by the executors of the Hofstra Estate. The idea of a college came up during a train ride between Penn Station and Hempstead. In 1935, the Hofstra College opened as an extension of New York University. The mansion house was the entire college that first year. A year later, Hofstra University became an independent institution and in 1963 it was given university status.

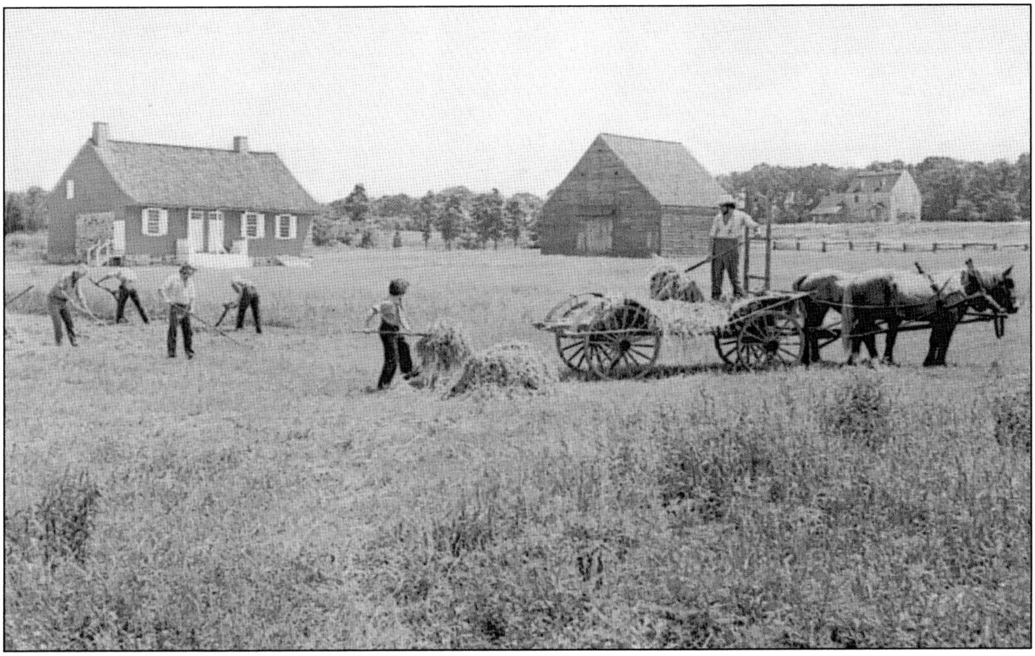

Bethpage Restoration Village is a pre-Civil War restored, countryside village containing authentic buildings and artifacts. Its 200 acres abound in representations of the lifestyle of English and Dutch settlers.

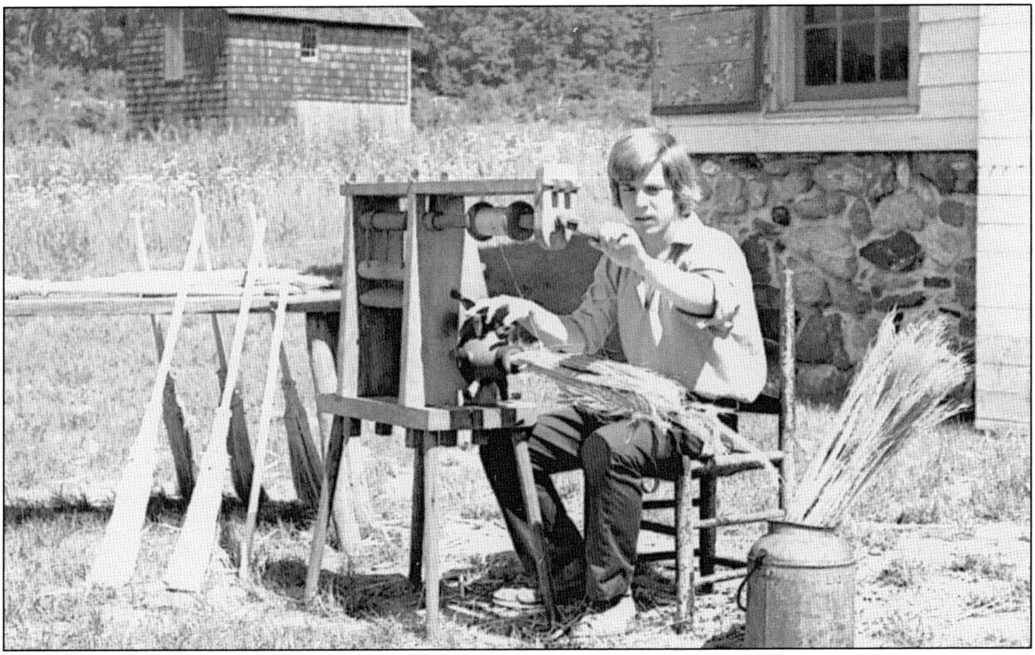

Bethpage Restoration Village features hands on experiences. Authentic craftsmen demonstrate their crafts and a working farm depicts the Civil War era. At the crossroads of this village, life centers on the inn, the general store, and the blacksmith shop. Here one may enjoy the presence of the past. Bethpage is the home of the Luna Module that landed on the moon in 1969. It was made at Bethpage's Grumman Plant in 1969.

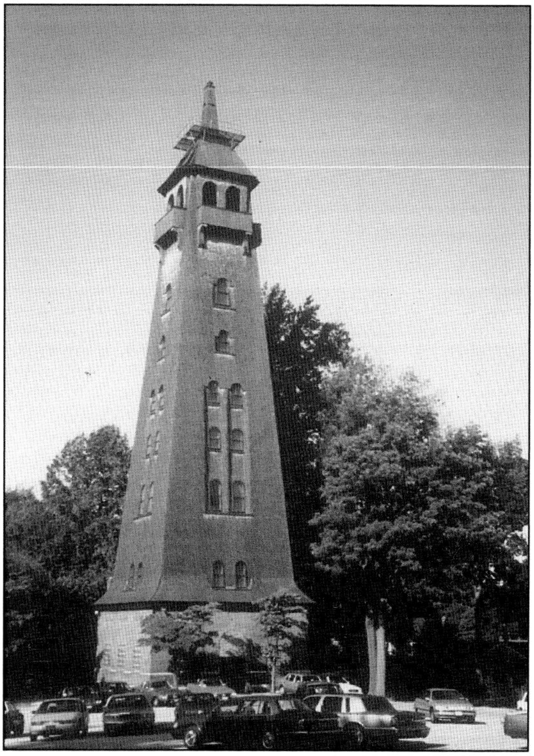

Whitney Windmill on Old Westbury Golf Course was built by William C. Whitney to provide water for his estate. Lightening in 1940 destroyed the sails. It is the most elaborate windmill to serve the Long Island water system. Whitney attained a monopoly and made a fortune on the New York Transportation system. He was the grandfather of Joan Payson who was instrumental in the establishment of the New York Mets baseball team.

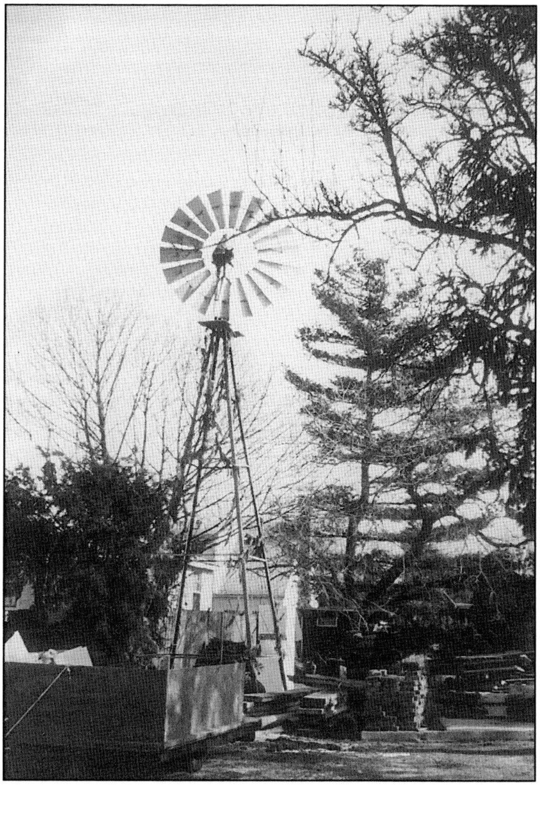

The windmill at Baldwin was Frank Davison's farm mill in its early days. Windmills have been in existence on both sides of the Atlantic for hundreds of years. They were essential to survival. Where waterpower was limited, Americans turned to wind. Colonists brought with them the millwrighting skills they needed. On Long Island, they relied heavily on the wind.

Alexander T. Stewart, an Irish-born merchant, founded Garden City in 1869. He was the owner of Wanamaker's Department Store and a tycoon in his field. Over 7,000 acres of land on the Hempstead plains were his domain. He built a hotel and railroad into his property. In 1883, his wife, Mrs. A. Steward, had the Cathedral of the Incarnation built in his memory. It contains 70 stained-glass windows depicting the Incarnation.

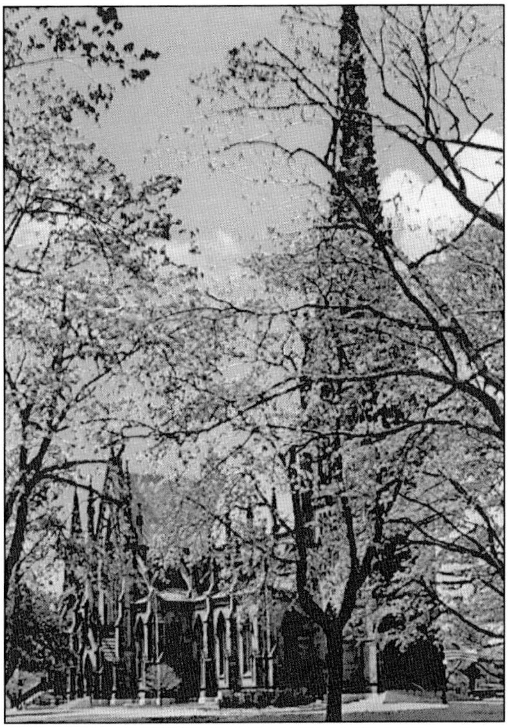

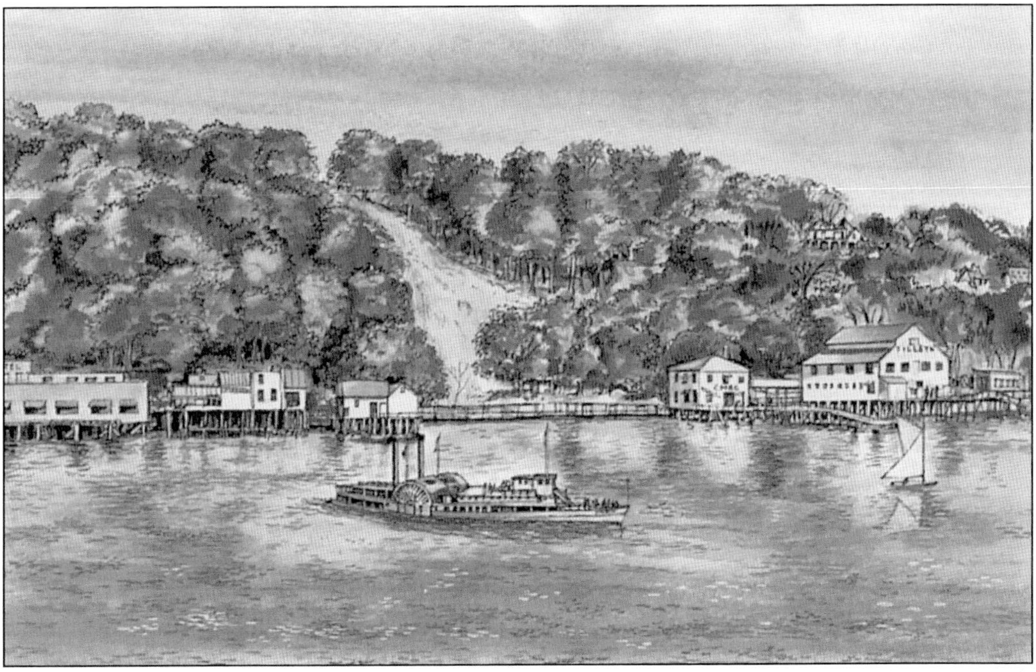

Sea Cliff is located on a bluff overlooking Long Island Sound. During the revolution British troops monitored water traffic on the sound. In the 1870s, the Metropolitan Methodist Campground Association purchased 250 acres for prayer and revival meetings in Sea Cliff. Lots were laid out, tents were set up, and the narrow pathways between the tents are the narrow roads of today. To facilitate transportation from cliff to shore a cable car was installed in 1886.

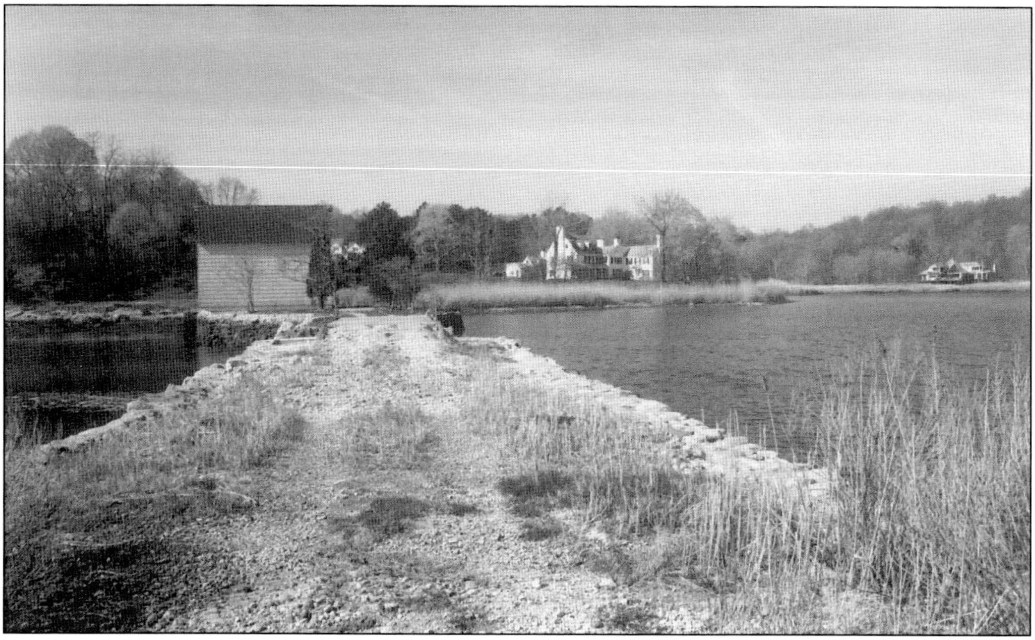

Lefferts-Van Wyck Tidal Mill was built in 1795 for Abraham Van Wyck, who owned an estate in Huntington. Historically significant, the mill is one of very few in the country that has remained on its original dam and has its original wooden gear intact. In later years, the Lefferts family bought the estate and mill.

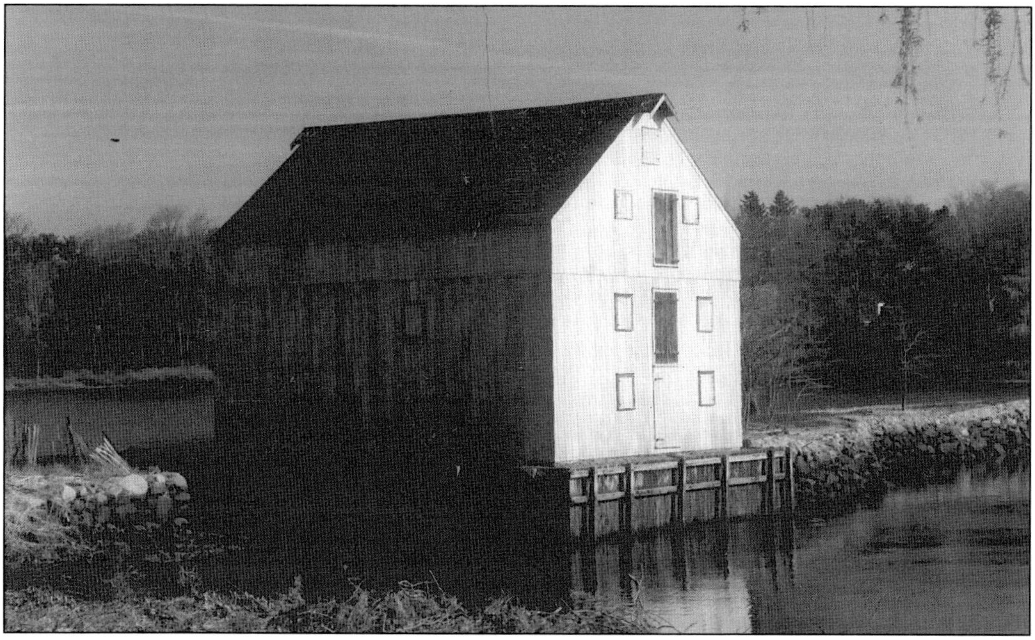

Lefferts bought the Van Wyck mill and worked it for many years. It was operational until 1893, when small mills declined because larger and more economical steam rollers appeared. The Industrial Revolution altered the farmers' need for mills. Leffers-Van Wyck Mill is now owned by the Nature Conservatory as part of its 16-acre Mill Cove Waterfowl Sanctuary. Lefferts' historic gristmill stands in Huntington's Lloyd Neck Harbor.

Huntington's historic windmill was the most conspicuous building in the village. An ingenious gentleman, Daniel Sammis, built it in 1825. He designed it as a sawmill. It was one of a kind with a 50 foot diameter. The young lads of the area enjoyed the exhilarating but dangerous sport of riding on its rims. Power was derived from its horizontal wheel that was suspended from a central shaft. There were 18 sails.

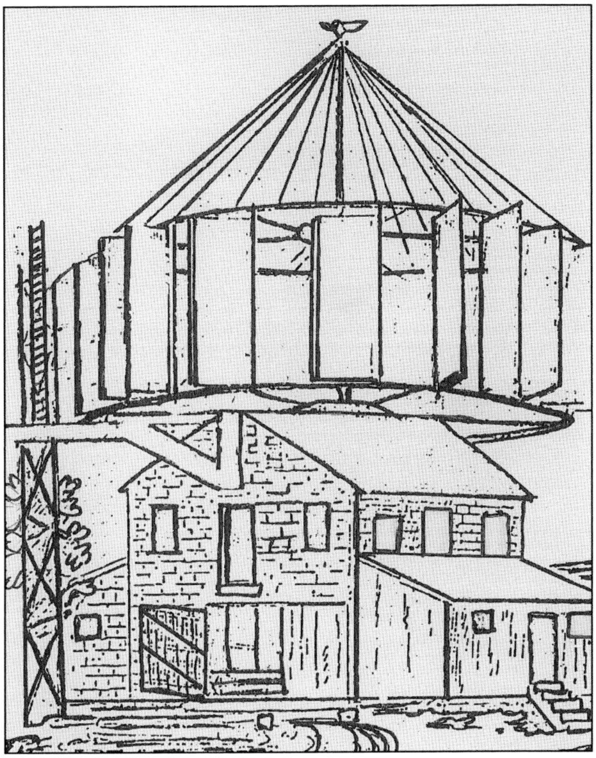

Cottages on the waterfront date back to the days when fishermen and salt hay seekers constructed crude, single, one-room, temporary dwellings to save time spent on the bay. These were made of driftwood or lumber from the mainland. They served as shelters in which to take refuge at day's end. When the era of waterfront recreational pursuits was ushered in, vacationers built more sophisticated cottages along the waterfront.

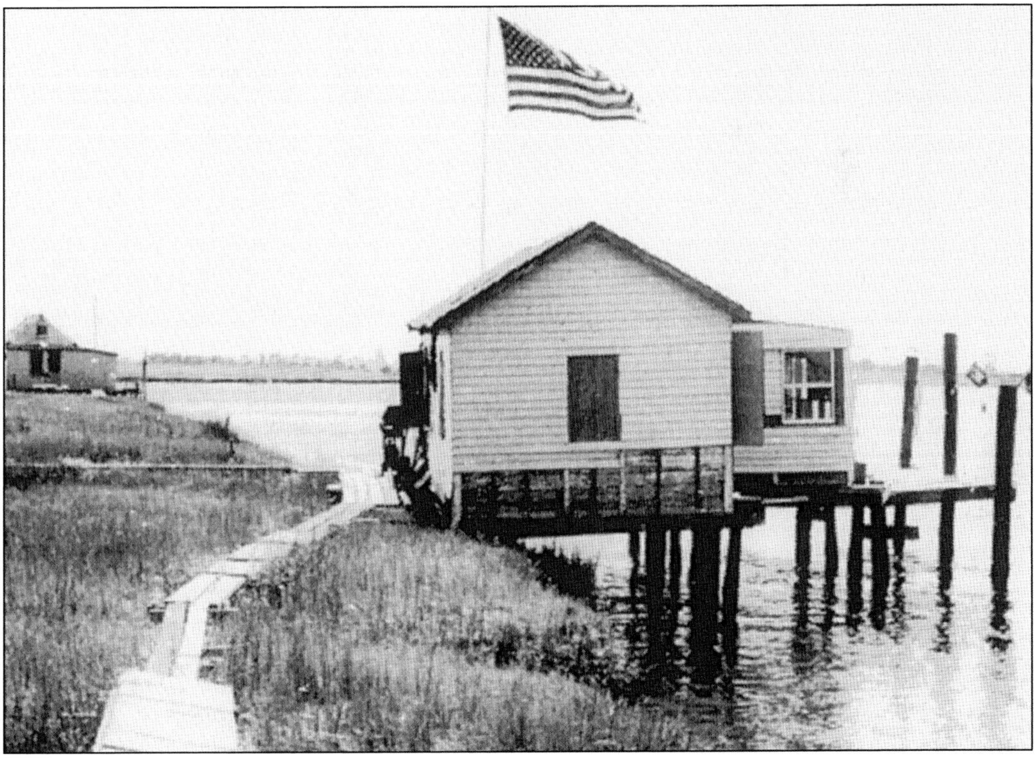

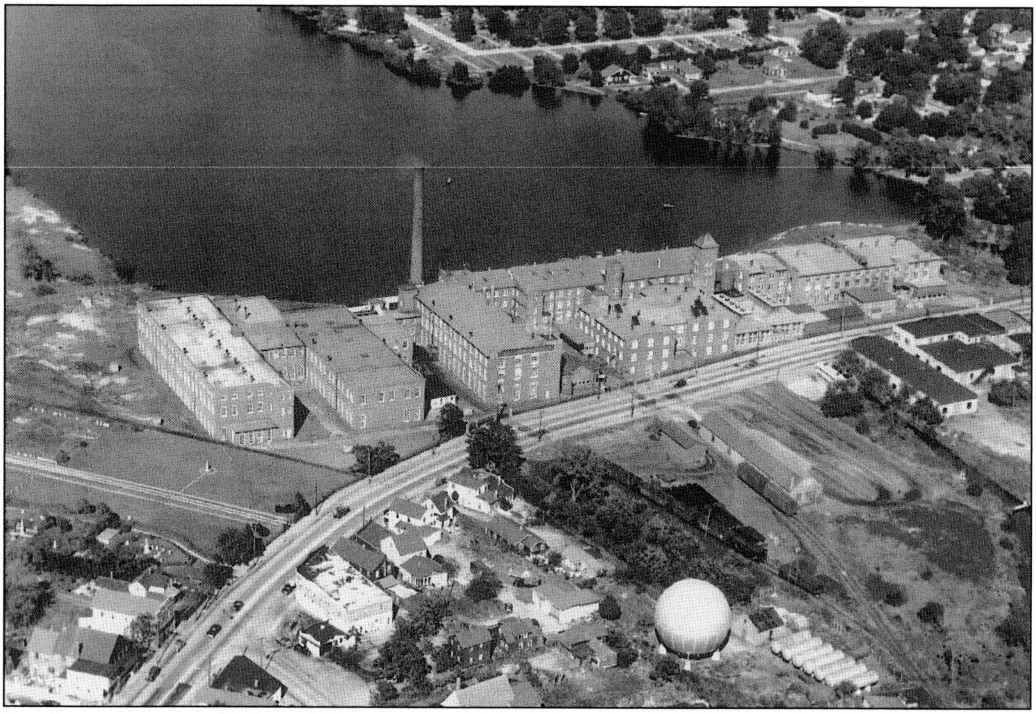

Patchogue Lace Mill dates back to 1890. Looms were installed and immigrants were hired from England and Italy as skilled weavers or laborers who worked in the bleaching and dying departments. In 1918, the Boston Co. took it over and called it Plymouth Mills. Later it was called Patchogue Plymouth Mills. The mill reached a pinnacle of success in 1930 with 900 employees from West Sayville to Center Moriches. The products were sold to department stores throughout the country.

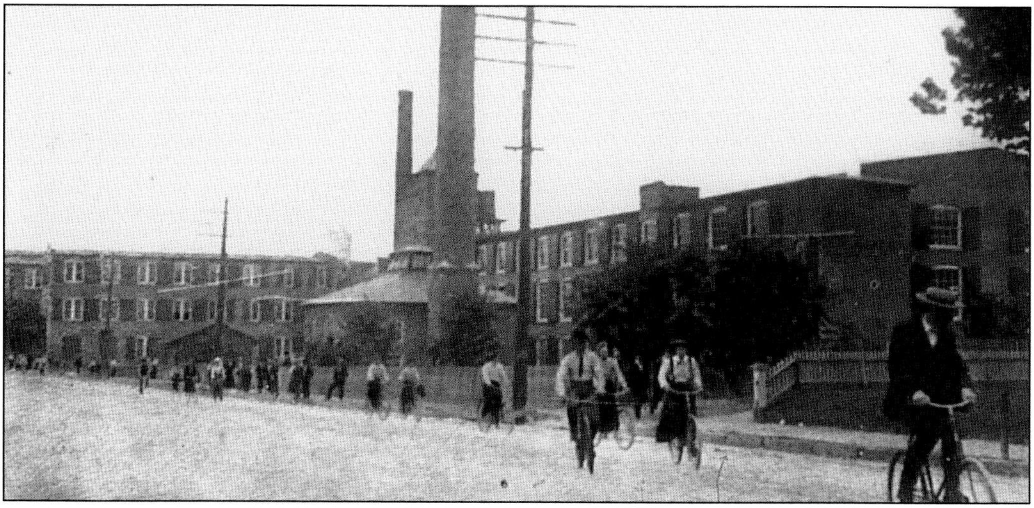

A change of shifts at the Patchogue Lace Mill sent employees to their bikes. This was in the heyday of the twenties when the mill was the mainstay of the community. The day revolved around the mill's whistle. Sometimes it was dubbed "The College" because seniors finishing high school could find employment at the mill. Many generations of families were employed at the mill.

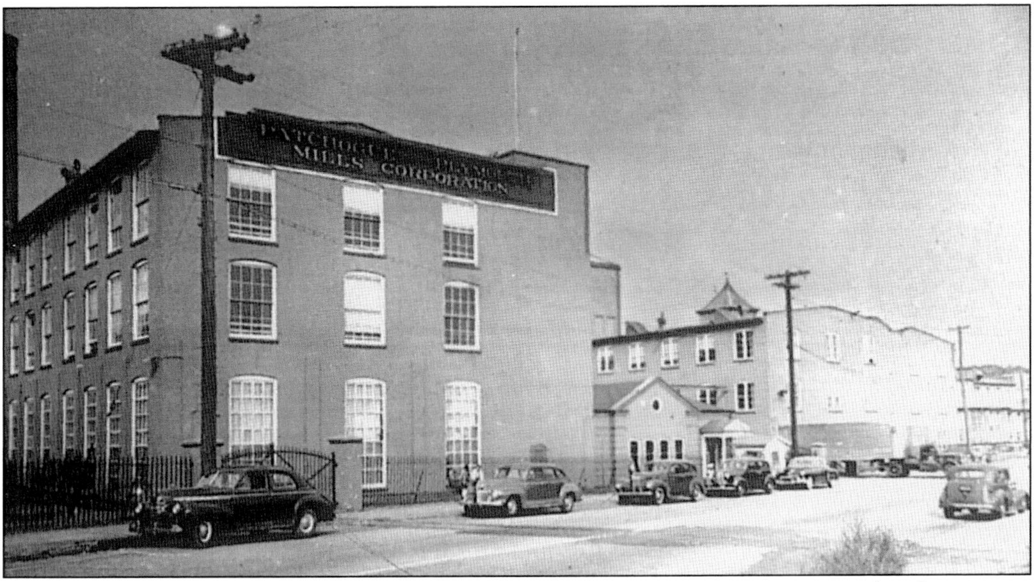

Patchogue was once the leading lace production center of the nation. Each of its many buildings housed a different aspect of the process. The brown room derived its name from the fact brownish material was sent to the mill. The washing and bleaching department quickly took care of that. Seamstresses and sewing machine operators depended on southern exposure for light and the scissors sharpening department was kept constantly busy as were administrators and bookkeepers.

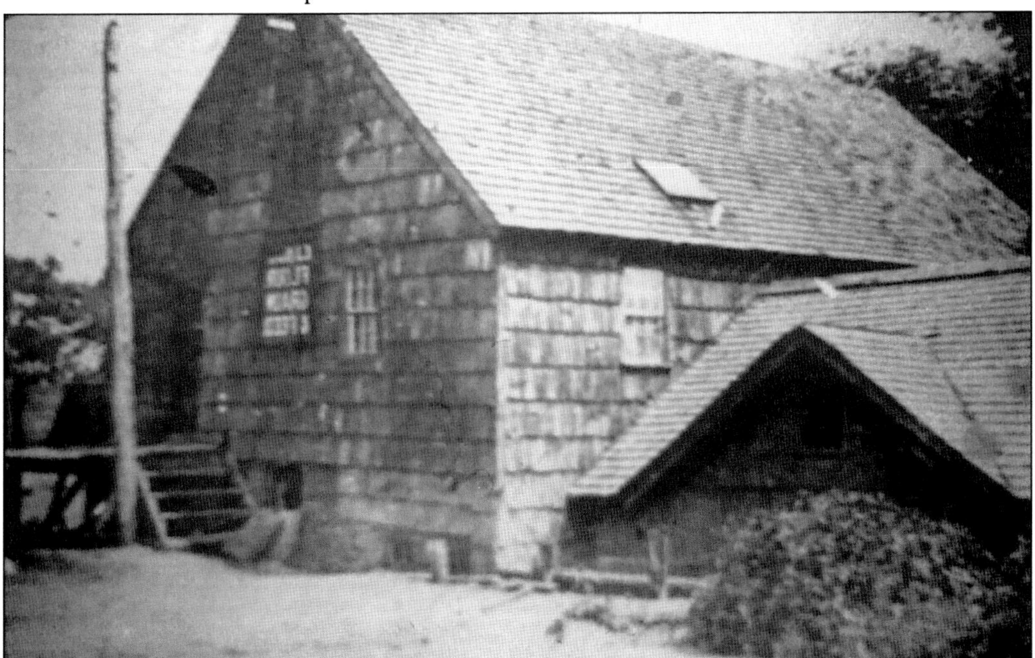

The Overton Grist Mill on Patchogue Lake preceded the lace mill. The exact date of its establishment is uncertain, but it was in operation between 1888 and 1900. Its second owner was Elbert Overton. The millstones are now part of Sweezey's Department Store. There was a millrace under the mill.

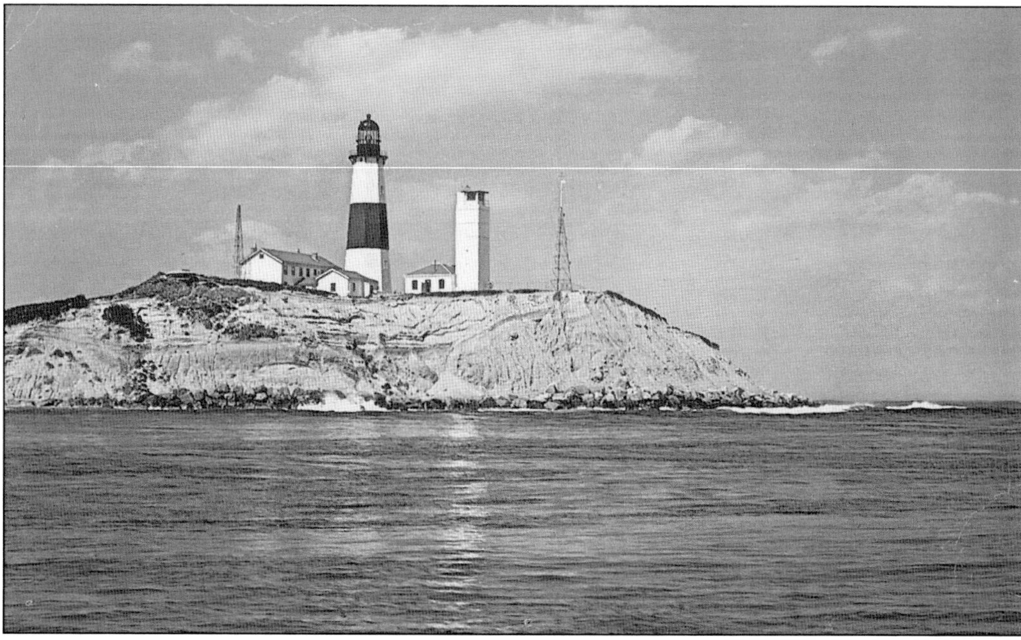

Montauk Lighthouse stands on a precipitous bluff marking the extreme tip of Long Island. The 100-foot octagonal tower was authorized by President George Washington and completed in 1797. Montauk is the oldest lighthouse in New York State. Iron steps—137 of them—take one to the tower. The lighthouse was automated in 1987 and now the light flashes every 5 seconds and is visible at 19 nautical miles. It is located 5 miles east of the quaint fishing village of Montauk.

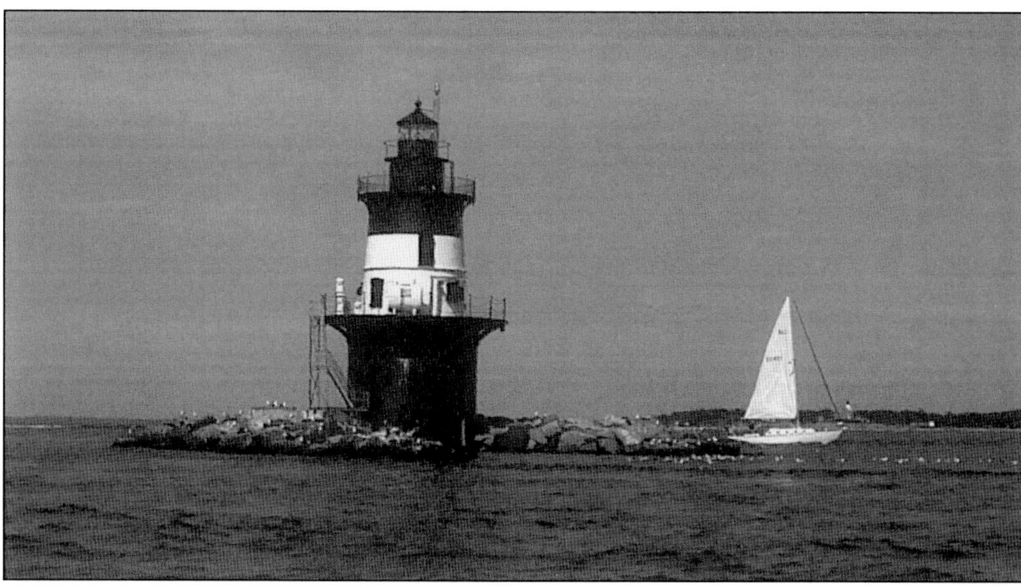

Orient Point Lighthouse serves harbors on the North Fork. It was completed in 1899, and the red beacon atop a 64-foot tower is visible at a 14-mile range. A three-man crew operated it until 1958, when it was automated. The lighthouse is located on Plum Gut between Long Island Sound and Gardiners Bay. Many yachts sail past it between Long Island and Connecticut.

The windmill in Bellport was located at the Gateway Playhouse. This was a converted barn where shows were periodically held. The playhouse remained opened for over three decades. Many interesting shows were produced here. Windmills are testimonials of a native technology.

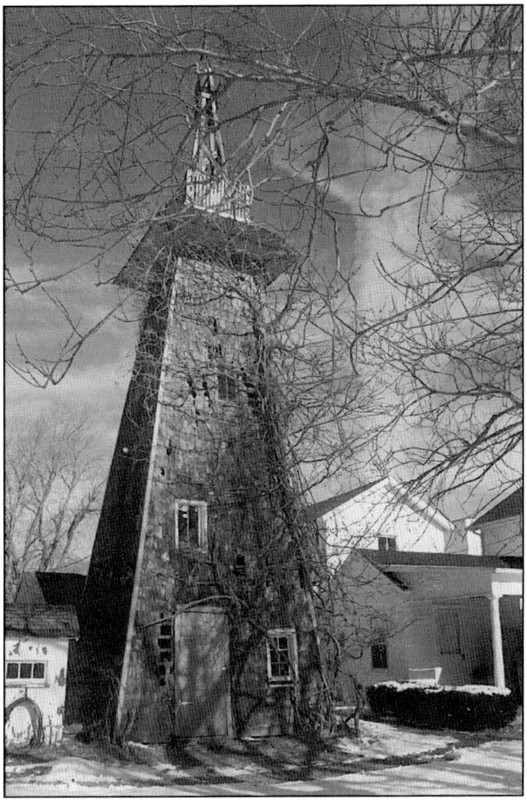

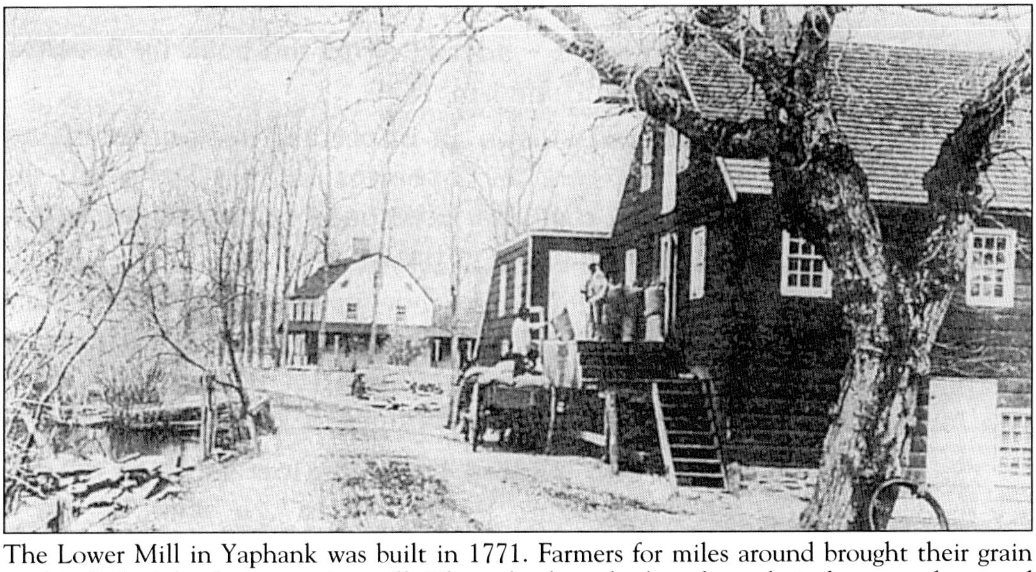

The Lower Mill in Yaphank was built in 1771. Farmers for miles around brought their grain to this little inland community mill. They also brought logs from their forests to be sawed into timber. The site of the windmill would later become Camp Upton. Later still Brookhaven National Laboratories would be housed here.

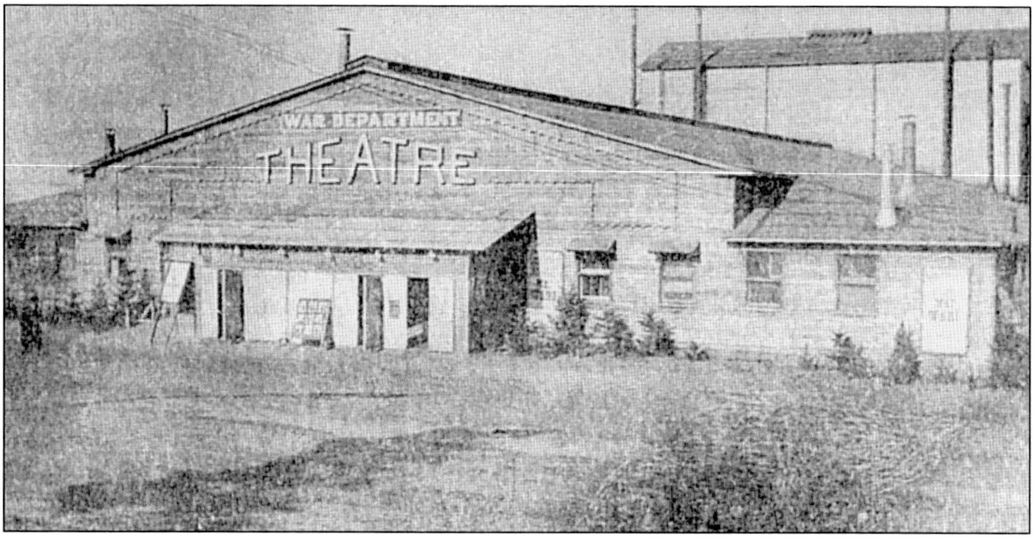

Camp Upton was established in 1917 when the federal government purchased 6,000 acres in Yaphank. Barracks were built and named after a Civil War general. Upton was an induction center that accommodated four thousand soldiers. It had its own water supply, sewerage system and fire department. Irving Berlin was among the soldiers. He wrote "Oh How I Hate To Get Up In The Morning" while at Upton and it was in the theater here that he performed "Yip Yap Yaphank."

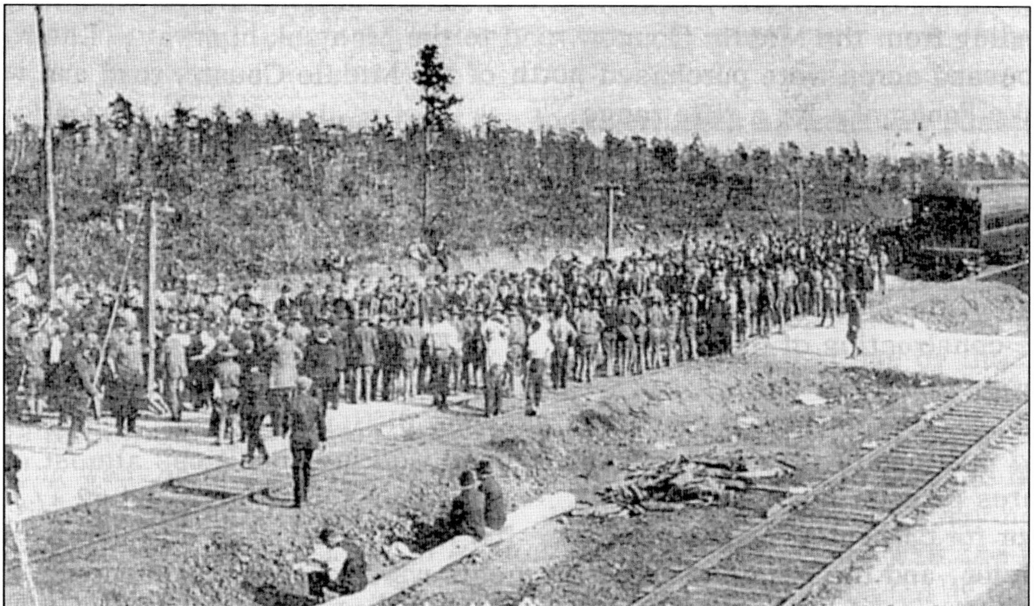

Thousands of men were trained at Camp Upton. It became a debarkation center for soldiers' discharge. On August 21, 1921, a public auction was held and the camp was to be demolished. Builders and developers came to obtain lumber and materials at their own price. Purchasers took the buildings down and salvaged the lumber. The 16,000 buildings and utilities had to be removed within 60 days. Purchasers relocated smaller buildings. Brookhaven National Laboratories was later established on this site.

Seven

NOSTALGIC PASTURES

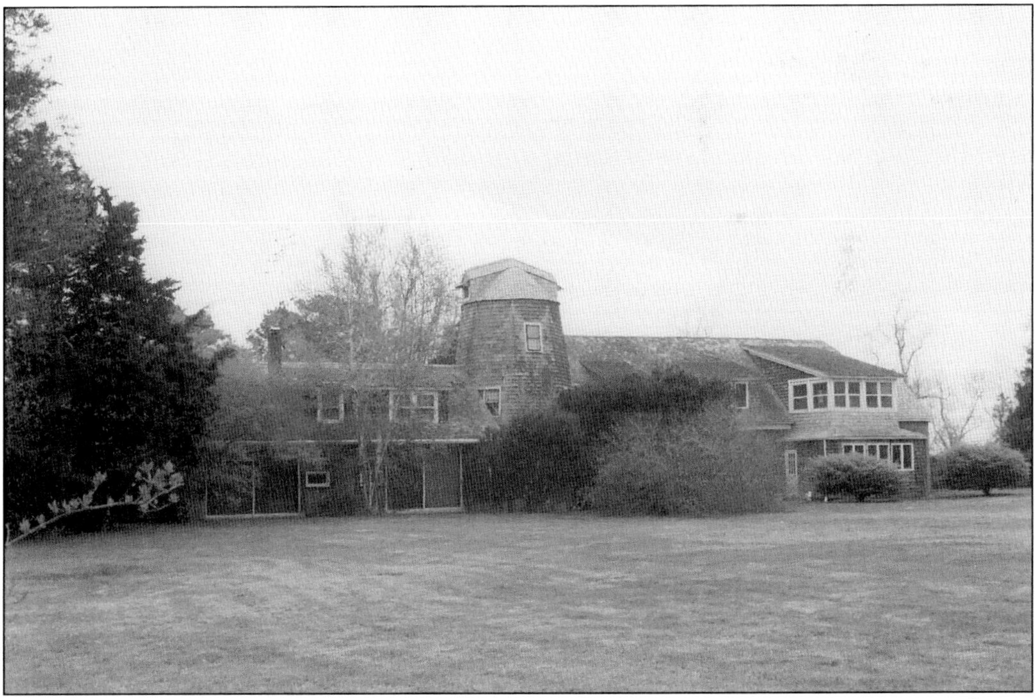

The Wainscott Mill is nestled between two homes. Windmills have long ago retired to a nostalgic pasture. No longer in operation they now dot the landscape as sentinels of yesteryear. Long Island windmills are unique in that they form the largest collection of historic windmills in the country. Some have been adopted by loving families who make them part of their homes. Such was the good fortune of a mill in Wainscott.

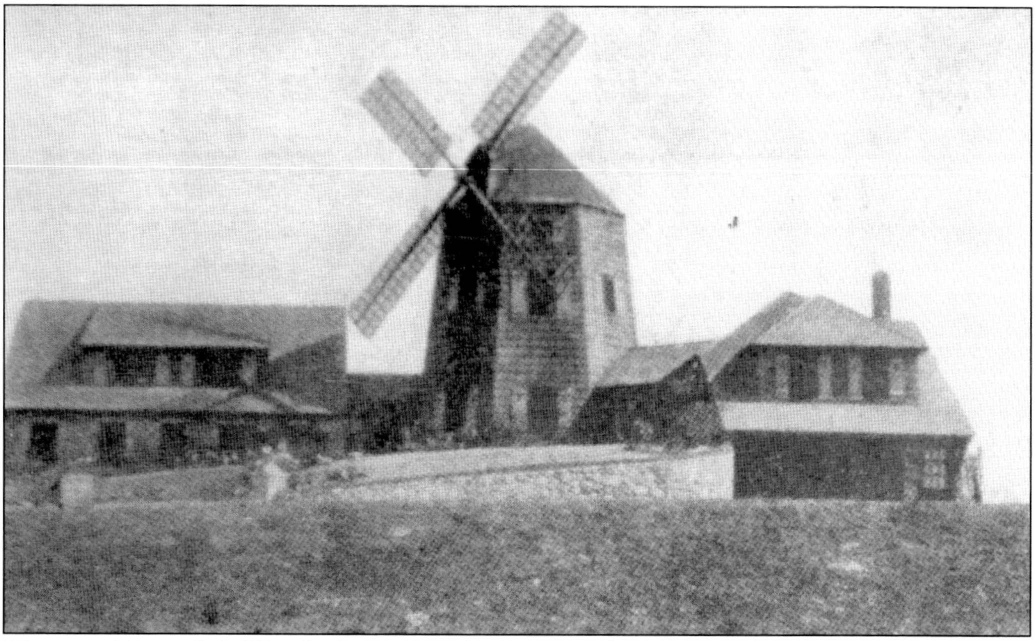

This was Montauk Mill, which sojourned on Montauk Point. High up on Sandpiper Hill, overlooking the Atlantic Ocean, the mill became part of the new home that Lathrop Brown was building. Eighteen years later it rejoined the walking mills when the Montauk bluffs became government property. The mill was relocated to Wainscott.

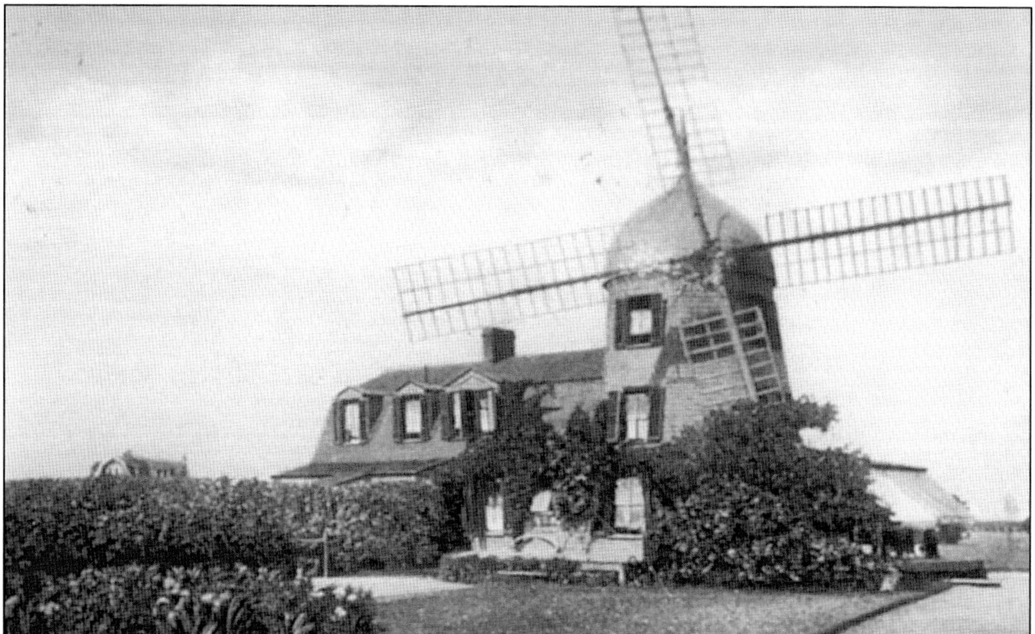

A picturesque windmill house in Southampton was one of many that dotted the eastern Long Island landscape as mills closed or were no longer needed. New technology took over the grinding tasks of the mills. Many mills were relocated to historic sights while being accepted as part of private homes.

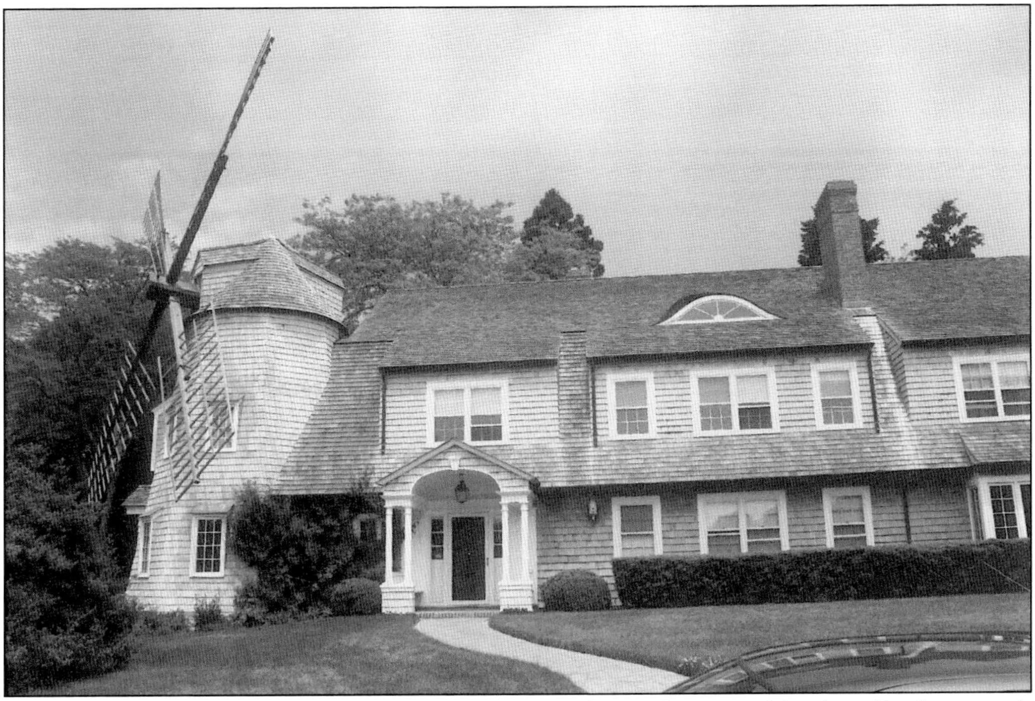

This house by the sea dates back to 1900. It stands on 3 acres of land in Easthampton's historic district. The main house, with its attached windmill, is remodeled as a contemporary dwelling. It has central air conditioning, a spacious library, and sun porch akin to its counterparts of yesteryear.

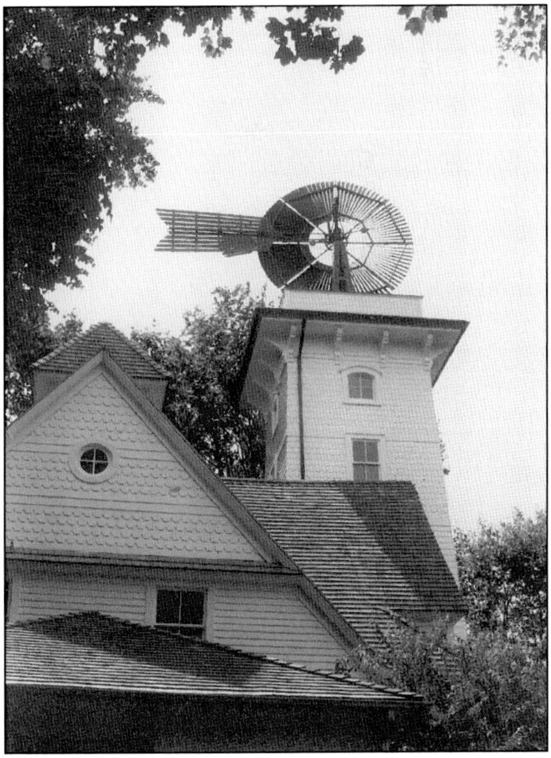

This retired mill on Shelter Island's Nostrand Avenue, became part of a dwelling. Windmills each have their own character and individuality. They represent a wooden technology that we cherish and preserve for future generations. These architectural treasures are the forerunners of our modern technology.

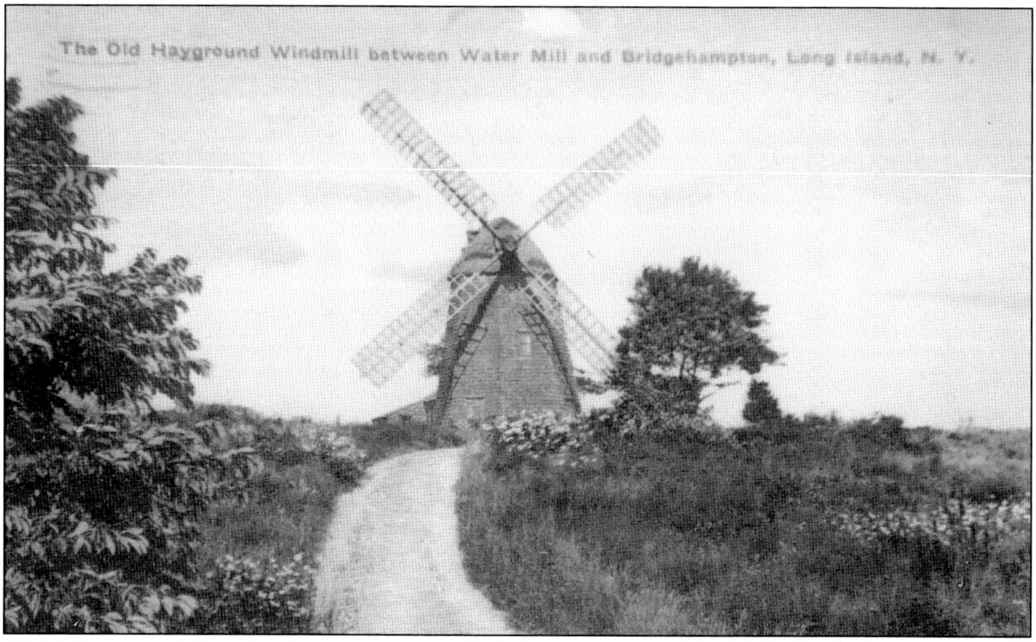

Old Hayground Windmill is in its primitive setting. Windmills have been regarded as picturesque remnants of colonial times. They provide a flashback to another era. The many mills that survive on eastern Long Island, the largest collection in the country, provide opportunity to investigate the evolution of technology. The mill is a very intensely practical, useful, and crowded piece of engineering.

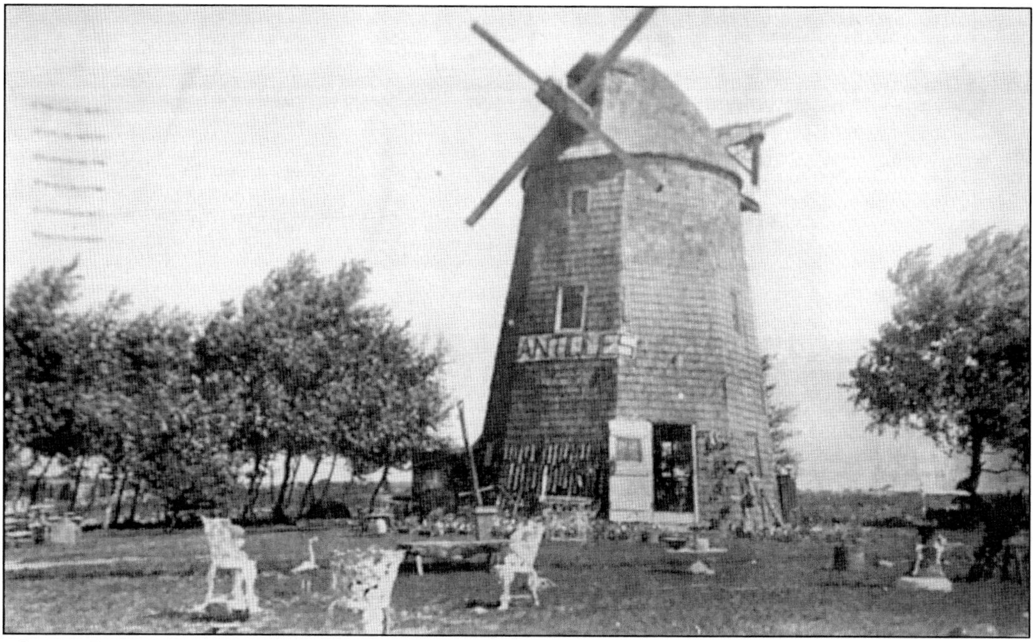

This mill near Bridgehampton has taken on a second career as an antique shop. Mills ceased operating when trains shipped flour to eastern Long Island. The area gradually became the potato capital of the east. It also became a resort area for artists, poets, and songwriters seeking inspiration and recreational pursuits.

William Sydney Mount (1807–1868) was one of four brothers born in Setauket. While serving as an apprentice to his brother, he began to draw. Later he traveled the countryside in a horse-drawn wagon that became his studio. He captured country scenes on canvas and became known as the "Painter of American Life." His mobile studio afforded him opportunity to paint Long Island scenes. The Metropolitan Museum of Art is one of many galleries that now hold his work.

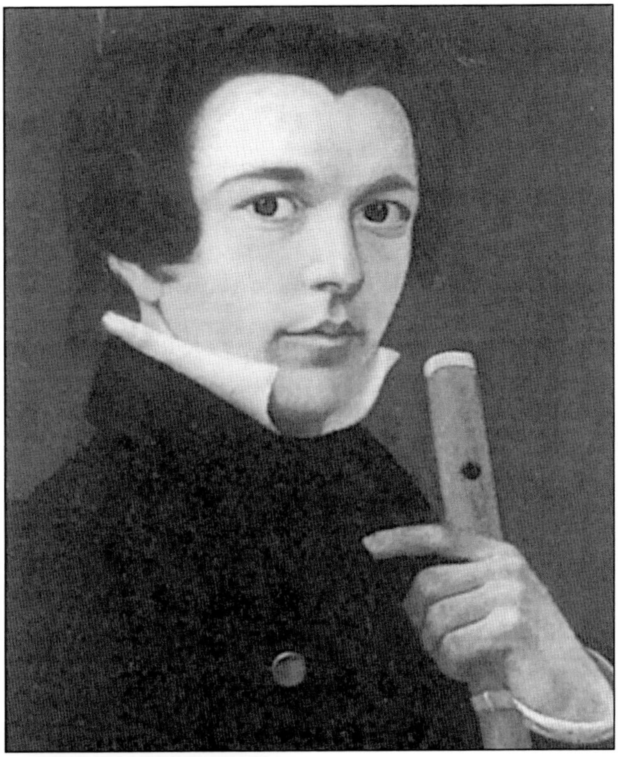

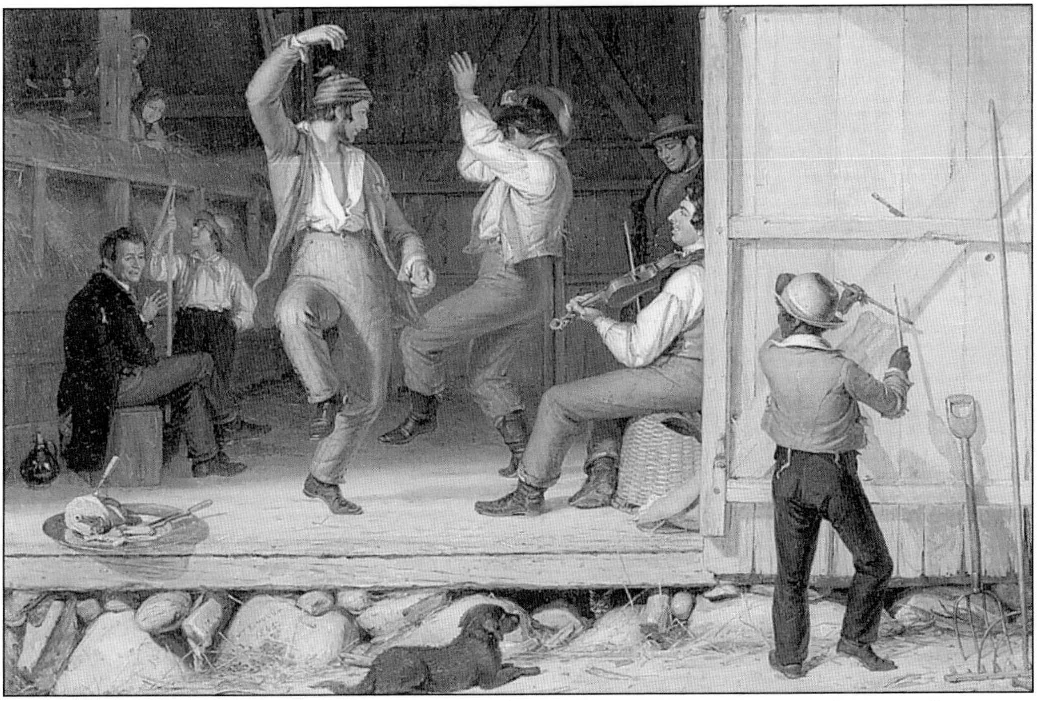

Dance of the Haymakers was one of William Sydney Mount's country scenes. Among others were *Dancing in the Barn*, *The Power of Music*, *Just In Tune*, and *The Banjo Player*. This is a portrayal of African Americans associated with the minstrel shows of the period.

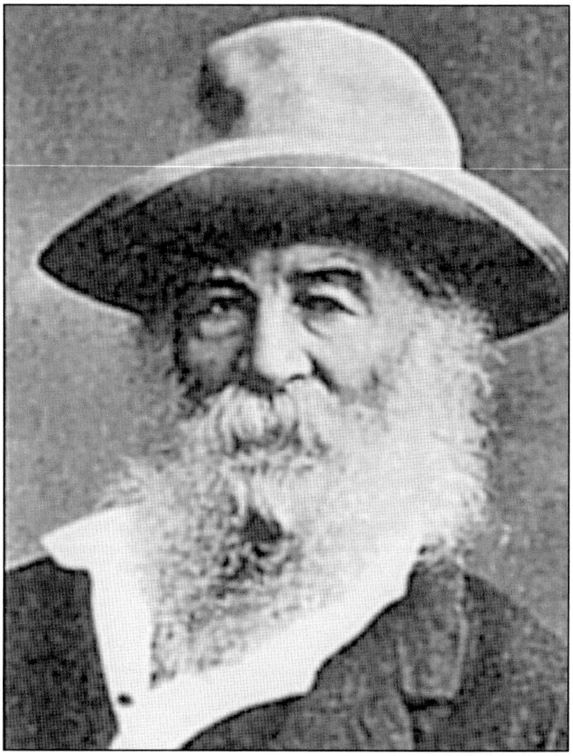

Walt Whitman was born in West Hills, near Huntington, in 1819. He always wrote well of his beloved Long Island. He founded the *Long Islander*, a weekly newspaper in Huntington. He taught school and edited newspapers, including the *Brooklyn Eagle*. The teaching profession took him to the Smithtown area. He became a printer, writer, and poet. His home, built in 1870, is now a museum on New York Avenue in Huntington. He is famous for his *Leaves of Grass*.

Huntington was the site in which Nathan Hale was captured by the British for spying during the Revolution. Dressed as a Dutch school teacher, Nathan had collected considerable documents when the British captured him and found the information in the sole of his shoe. He was hanged in New York City. His response was: "I regret I have but one life to give for my country."

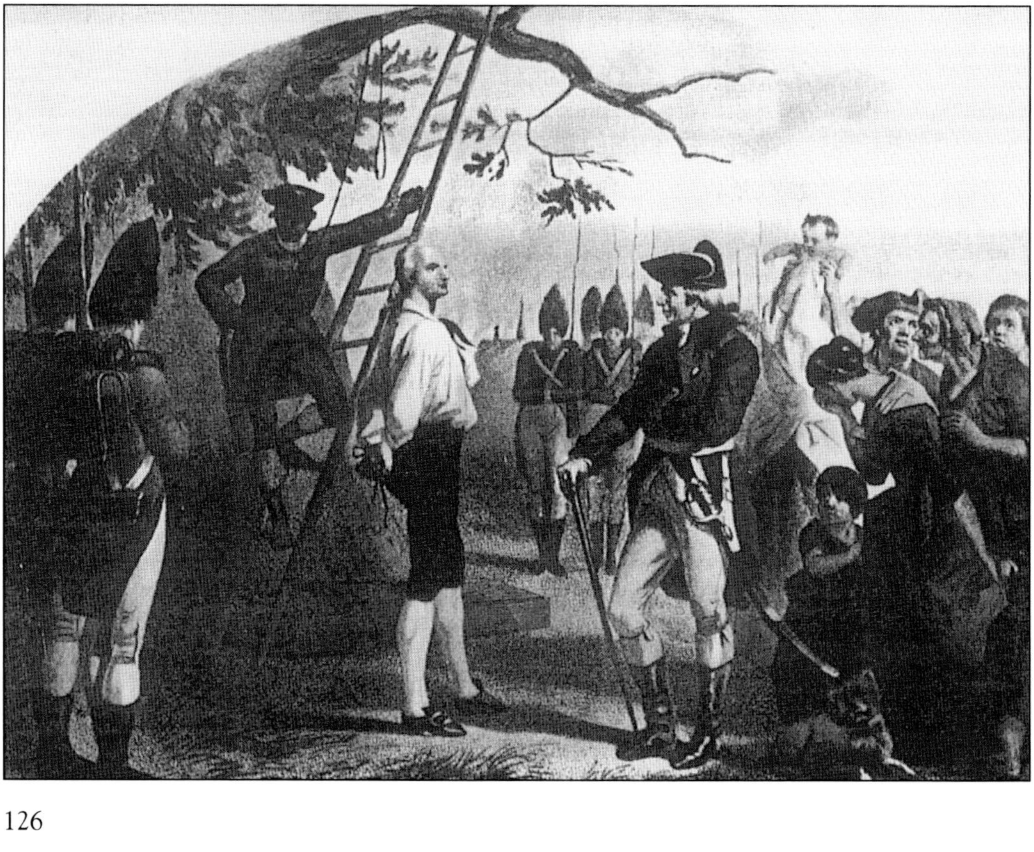

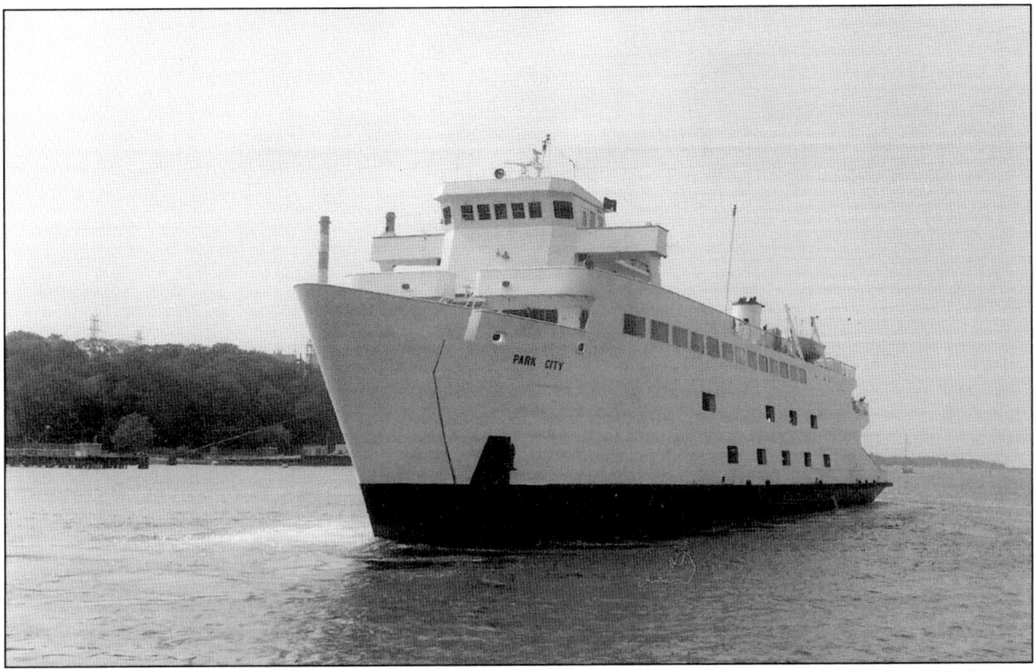

The *Grand Republic* ferry crosses Long Island Sound on excursion trips daily. Port Jefferson is a crescent shaped harbor on the North Shore. John Paul Jones had rendezvous with his Revolutionary fleet here. The colonial settlers dubbed it Drowned Meadow to discourage commercialism. Pioneers agreed it was too breathtaking to be trampled by crowds. They called it Mount Misery. In 1860, the picturesque little village became Port Jefferson. Today the village houses the Bridgeport Ferry.

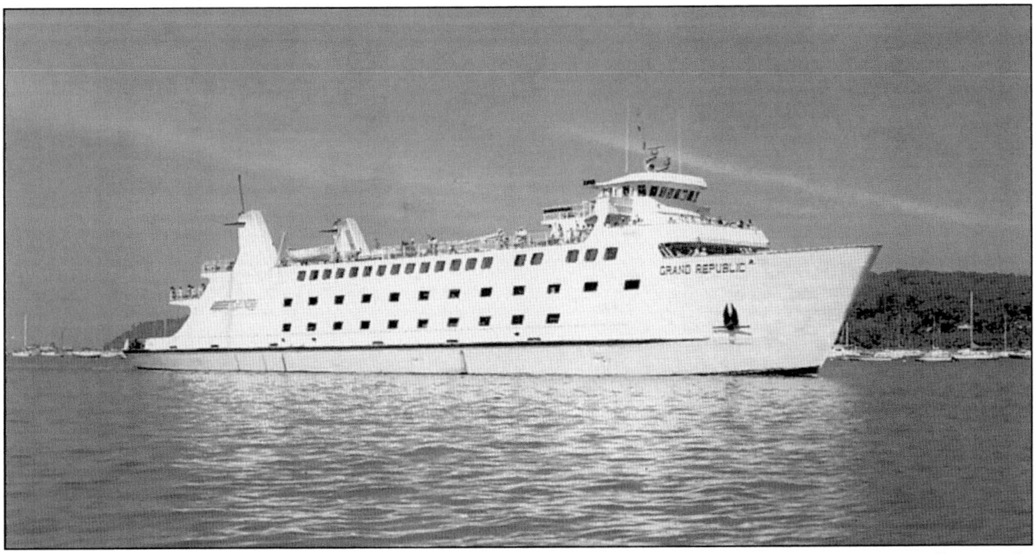

Ferry service across Long Island Sound has been in existence for centuries. It was the mode of transportation for many of our early settlers from New England. In recent years, this ferry service has been established between Port Jefferson and Bridgeport, Connecticut, on daily sailing excursions. The Park City Ferry is part of the fleets that ply the waters of the sound and transport millions between the two ports.

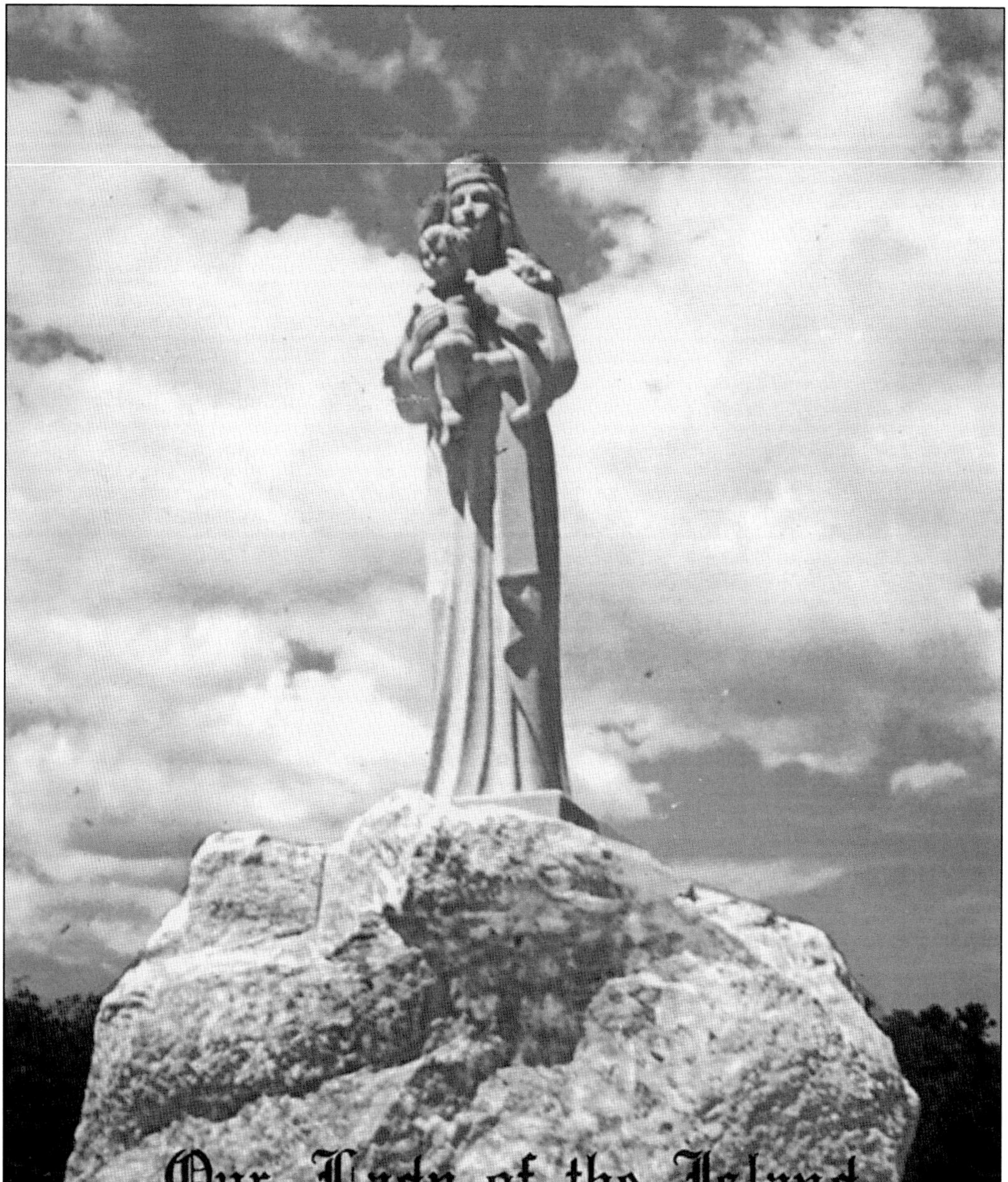

The Shrine of Our Lady of the Island was established at Eastport, Long Island, in 1975. The idea was conceived two decades earlier at the Montford Missionary Center in Bayshore. The humble little plan evolved into a magnificent shrine with an 18-foot statue of Mary, on a natural boulder, that towers over the Atlantic Ocean. Crescenzo Vigilotta donated 60 acres in Eastport and the great boulder and the Harrison family donated land on which it rests. Rev. William Vigliotta of the Montford Fathers was the first curator.

Set in the peaceful stillness of wooded hillsides, the Stations of the Cross are connected by a winding footpath. Life-size figures recreate the sacred settings. Over 150 juniper bushes, spaced in a quarter mile circle, count off the prayers of the rosary. Bishop John McGann dedicated the shrine on October 2, 1976.